X Workshop of the Gravitation and Mathematical Physics Division, Mexican Physical Society 2013

Journal of Physics: Conference Series Volume 545

Hidalgo, Mexico
2 – 6 December 2013

Editors:

R. B. Barcenas
H. H. Hernandez Hernandez
M. Sabido
C. A. Soto
L. A. Lopez
O. Pedraza
V. Ceron

ISBN: 978-1-63439-660-8
ISSN: 1742-6588

Printed from e-media with permission by:

Curran Associates, Inc.
57 Morehouse Lane
Red Hook, NY 12571

Some format issues inherent in the e-media version may also appear in this print version.

Printed by Curran Associates, Inc. (2015)

For permission requests, please contact the Institute of Physics
at the address below.

Institute of Physics
Dirac House, Temple Back
Bristol BS1 6BE UK

Phone: 44 1 17 929 7481
Fax: 44 1 17 920 0979

techtracking@iop.org

Additional copies of this publication are available from:

Curran Associates, Inc.
57 Morehouse Lane
Red Hook, NY 12571 USA
Phone: 845-758-0400
Fax: 845-758-2634
Email: curran@proceedings.com
Web: www.proceedings.com

TABLE OF CONTENTS

Journal of Physics: Conference Series **545** (2014) 011001 doi:10.1088/1742-6596/545/1/011001

Preface: X Workshop of the Gravitation and Mathematical Physics Division of the Mexican Physical Society

The collection of papers in this volume was presented during the X Workshop of the Gravitation and Mathematical Physics Division of the Mexican Physical Society (DGFM-SMF), which was held in Pachuca, Hidalgo, México, December 2-6, 2013. The Workshop is a bi-annual series of conferences sponsored by the DGFM-SMF that started in 1993 with the purposes of discussing and exchanging the research and experience of the gravitational and mathematical physics communities in Mexico. Each Mexican Workshop has been devoted to subjects of broad interest, so that students, in particular, can have access to specialized courses and talks that allow them to raise up their qualifications as professional researchers. Recurrent topics in the Mexican Workshop are supergravity, branes, black holes, the early Universe, observational cosmology, quantum gravity and cosmology and numerical relativity.
Following our previous Workshops, distinguished researchers in the field, working in Mexico, were invited to give courses, whereas young researchers were invited for plenary lectures. More specialized talks were also presented in parallel sessions, with ample participation of researchers, and graduate and undergraduate students; most of the presentations have been included in these proceedings.
The contributions in this volume have been peer-reviewed, and they represent most of the courses, plenary talks and contributed talks presented during our Workshop. We are indebted to the contributors of these proceedings, as well as to the other participants and organizers, all for making the event a complete success. We acknowledge the professionalism of our reviewers, who helped us to keep high quality standards in all manuscripts.

Acknowledgments

The organizing committee would like to acknowledge the financial support of the Mexican National Science and Technology Council (CONACyT), the Mexican Physical Society (SMF), as well as several Institutions including: Centro de Investigación y Estudios Avanzados (CINVESTAV), Universidad Autónoma Metropolitana Iztapalapa (UAM-I), Universidad de Guanajuato (UG), Universidad Nacional Autónoma de México (UNAM). We especially acknowledge the support and the kind hospitality of our host, the Universidad Autónoma del Estado de Hidalgo.

Ricardo Becerril Bárcenas, Héctor H. Hernández Hernández, Miguel Sabido, Carlos A. Soto, Luis Alberto López, Omar Pedraza, Victoria E. Cerón.

Editors

Journal of Physics: Conference Series **545** (2014) 011002 doi:10.1088/1742-6596/545/1/011002

Peer review statement

All papers published in this volume of *Journal of Physics: Conference Series* have been peer reviewed through processes administered by the proceedings Editors. Reviews were conducted by expert referees to the professional and scientific standards expected of a proceedings journal published by IOP Publishing.

Journal of Physics: Conference Series **545** (2014) 012001 doi:10.1088/1742-6596/545/1/012001

The evolution of a spatially homogeneous and isotropic universe filled with a collisionless gas

Francisco Astorga[1], **Olivier Sarbach**[1,2] **and Thomas Zannias**[1]

[1] Instituto de Física y Matemáticas, Universidad Michoacana de San Nicolás de Hidalgo Edificio C-3, Ciudad Universitaria, 58040 Morelia, Michoacán, México.
[2] Gravitational Physics, Faculty of Physics, University of Vienna, Boltzmanngasse 5, 1090 Vienna, Austria.

E-mail: fastorga@ifm.umich.mx, sarbach@ifm.umich.mx, zannias@ifm.umich.mx

Abstract. We review the evolution of a spatially homogeneous and isotropic universe described by a Friedmann-Robertson-Walker spacetime filled with a collisionless, neutral, simple, massive gas. The gas is described by a one-particle distribution function which satisfies the Liouville equation and is assumed to be homogeneous and isotropic. Making use of the isometries of the spacetime, we define precisely the homogeneity and isotropicity property of the distribution function, and based on this definition we give a concise derivation of the most general family of such distribution functions. For this family, we construct the particle current density and the stress-energy tensor and consider the coupled Einstein-Liouville system of equations. We find that as long as the distribution function is collisionless, homogenous and isotropic, the evolution of a Friedmann-Robertson-Walker universe exhibits a singular origin. Its future development depends upon the curvature of the spatial sections: spatially flat or hyperboloid universes expand forever and this expansion dilutes the energy density and pressure of the gas, while a universe with compact spherical sections reaches a maximal expansion, after which it reverses its motion and recollapses to a final crunch singularity where the energy density and isotropic pressure diverge. Finally, we analyze the evolution of the universe filled with the collisionless gas once a cosmological constant is included.

1. Introduction

It has been recognized long ago that the relativistic kinetic theory of gases plays a vital role in the description of important processes in relativistic astrophysics and cosmology. Although the phenomenological description of matter by fluids or magneto-fluids seems to be adequate for many astrophysical or cosmological scenarios, it fails to provide a reliable description for some important cases. For instance, the interaction of electromagnetic radiation with a a tenuous atmosphere requires both media to be treated by methods of kinetic theory. Similarly, during a core-collapse supernova explosion, the propagation of neutrinos in regions where their mean free path becomes comparable or larger than the core dimensions, methods of relativistic kinetic theory are required. In cosmology, the general relativistic Boltzmann equation is employed for the description of photons, neutrinos and often electrons during or before the recombination era. Calculations of the cosmological helium production treats the participating species of elementary particles via distribution functions satisfying the relativistic Boltzmann equation. Nowadays relativistic kinetic theory has become a respectable, challenging field in mathematical relativity. The Einstein-Liouville and the Einstein-Boltzmann equations are relevant and on the frontier of

Journal of Physics: Conference Series **545** (2014) 012001 doi:10.1088/1742-6596/545/1/012001

studies in mathematical relativity, see for instance Refs. [1, 2, 3, 4, 5, 6] and Ref. [7] for a recent review.

Early studies of relativistic kinetic theory started immediately after the birth of special relativity. Work by Synge [8], Taubes and Weinberg [9], Israel [10], Lindquist [11], Ehlers [12, 13] and others led to the foundations for the modern description of the theory. In [14, 15, 16], inspired by these early approaches, we presented an introduction to relativistic kinetic theory that relies on the geometric and symplectic properties of the tangent bundle of the background spacetime manifold. In [14], the Hamiltonian structure of the theory has been highlighted, while in [15] the importance of the Sasaki metric on the tangent bundle for the physical interpretation of the theory has been emphasized. As we have shown, the Sasaki metric is particularly useful to introduce a naturally defined integration theory on the tangent bundle which is free of ambiguities, and it is also useful in order to interpret symmetries of the distribution function.

Making use of these additional insights in the structure of the theory, in [15] we also derived the most general spherically symmetric distribution function on an arbitrary spherically symmetric spacetime and the most general collisionless distribution function on a Kerr black hole background. In [16] we extended these results to the derivation of a collisionless distribution function describing a charged gas on a Kerr-Newman black hole background.

While our earlier work focused on the formal structure of relativistic kinetic theory and astrophysical applications, in this work we discuss the propagation of a simple, collisionless gas from a cosmological perspective. At first, we discuss the properties of a neutral, simple, massive gas (that is, a collection of neutral, spinless classical particles of the same positive rest mass) propagating on a spatially homogenous and isotropic Friedmann-Robertson-Walker (FRW) spacetime, and we derive the most general distribution function which shares the same symmetries as those of the FRW spacetime metric. As it turns out, such a distribution function has a very simple and tractable form: it is described by a function of proper time τ of the isotropic observers and a particular quantity C which is constructed from the integrals of motion associated to the Killing vector fields of the spacetime. For the particular case of a *collisionless*, spatially homogeneous and isotropic gas, the distribution function is required to satisfy the Liouville equation, and in this case it reduces to a function of the single quantity C, in accordance to previous results by Maharaj and Maartens [17, 18, 19]. For early applications of relativistic kinetic theory to cosmology see also [20, 21, 22, 23]. In particular, in Ref. [22], two relativistic collisionless gases are employed with the first one describing the population of normal galaxies and the other one the population of quasi-stellar objects. Their distribution function is chosen such that it maximizes an entropy functional, and the phenomenology of the model is analyzed in detail.

In a next step, we discuss the particle current density and stress-energy tensor of the gas. As long as the gas is described by a spatially homogeneous and isotropic distribution function, the particle current is parallel to the four-velocity of the isotropic observers while the stress-energy tensor has the form of a perfect fluid whose four-velocity (the dynamical mean velocity of the gas) is parallel to the particle current. However, the corresponding energy density ρ and isotropic pressure P are not arbitrary. Rather, they are given by fiber integrals of the distribution function over the future mass hyperboloid. Both, the density and pressure, are positive definite and the stress-energy tensor satisfies the weak, strong and dominant energy conditions. A particular interesting feature of the gas is that it behaves like a radiation fluid ($P \simeq \rho/3$) in the early universe, when the scale factor is very small, while in the late universe it behaves similarly to dust ($P \simeq 0$) [6].

In the last part of this article, we use the stress-energy tensor in order to describe the self-gravity of the gas through Einstein's field equations, and we discuss a few qualitative properties of the spatially homogeneous and isotropic solutions of the Einstein-Liouville system. We show that spatially flat or hyperboloid universes expand forever, diluting the energy density and pressure

Journal of Physics: Conference Series **545** (2014) 012001 doi:10.1088/1742-6596/545/1/012001

of the gas, while a universe with compact spherical sections reaches a maximal volume, after which it reverses its motion and recollapses to a final crunch singularity.

Motivated by the recent results coming from the analysis of distant supernovae suggesting that the universe is currently in an accelerated state of expansion [24, 25], we also discuss the case where a positive cosmological constant $\Lambda > 0$ is included. For spatially flat or hyperboloid universes there is again an everlasting expansion, but in contrast to the previous case with $\Lambda = 0$ where the expansion is damped due to the gravitational pull, now there is an accelerated expansion in the final stage of the evolution. For compact spherical sections the universe may expand forever or recollapse, depending on the values of the cosmological parameters.

Limitations of our approach and future work are discussed in the conclusions.

2. Spatially homogeneous and isotropic collisionless distribution functions

In this section, we consider a fixed FRW spacetime, and we review its symmetry properties and the derivation of the most general distribution function sharing these symmetries. Our derivation is based on the recent work in [15], where a systematic discussion on symmetries of distribution functions is provided. More specifically, it has been shown that infinitesimal generators ξ of one-parameter groups of isometries of the base manifold (M, g) lift naturally to infinitesimal generators $\hat{\xi}$ of one-parameter groups of isometries on the tangent bundle $(TM, \hat{g})$, where $\hat{g}$ is the Sasaki metric which is naturally induced from the spacetime metric g. Furthermore, it has been shown in [15] that the lifted generator $\hat{\xi}$ is tangent to the mass shells $\Gamma_m := \{(x,p) \in TM : g_x(p,p) = -m^2\}$ associated to particles of rest mass m and that it commutes with the Liouville vector field L: $[L, \hat{\xi}] = 0$. A distribution function f on a particular mass shell Γ_m is said to be invariant with respect to the isometry generated by $\hat{\xi}$ if the Lie derivative of f with respect to $\hat{\xi}$ vanishes, $\pounds_{\hat{\xi}} f = 0$. Accordingly, if the spacetime manifold admits a Lie group G of isometries, generated by Killing vector fields $\xi_1, \xi_2, \dots, \xi_r$, the distribution function is said to be G-symmetric if $\pounds_{\hat{\xi}_a} f = 0$, $a = 1, 2, \dots, r$.

2.1. *Killing algebra of the FRW spacetime*

In the case under consideration in this article the spacetime (M, g) has the form $M = (0, \infty) \times \Sigma_k$ and

$$g = -d\tau^2 + a(\tau)^2 \frac{dx^2 + dy^2 + dz^2}{\left(1 + \frac{k}{4}|\underline{x}|^2\right)^2}, \qquad |\underline{x}| := \sqrt{x^2 + y^2 + z^2}, \tag{1}$$

where $a(\tau)$ is the scale factor, and Σ_k is either hyperbolic space H^3, Euclidean space E^3 or spherical space S^3 depending on the (normalized) value of the constant spatial curvature $k = -1$, 0 or 1. The metric (1) is spatially homogeneous and isotropic, and the symmetry group G is generated by the following particular Killing vector fields of (M, g):

$$\xi(\underline{s}) := (\underline{s} \wedge \underline{x}) \cdot \frac{\partial}{\partial \underline{x}}, \tag{2}$$

$$\eta(\underline{b}) := \left(1 - \frac{k}{4}|\underline{x}|^2\right) \underline{b} \cdot \frac{\partial}{\partial \underline{x}} + \frac{k}{2}(\underline{b} \cdot \underline{x})\underline{x} \cdot \frac{\partial}{\partial \underline{x}}, \tag{3}$$

where $\underline{s}, \underline{b} \in \mathbb{R}^3$ are constant vectors, $\underline{x} := (x, y, z)$, and the wedge and the dot denote the standard vector and scalar products in $\mathbb{R}^3$. The Killing vector fields $\xi(\underline{s})$ and $\eta(\underline{b})$ satisfy the following commutation relations for all $\underline{s}_1, \underline{s}_2, \underline{b}_1, \underline{b}_2 \in \mathbb{R}^3$:

$$[\xi(\underline{s}_1), \xi(\underline{s}_2)] = -\xi(\underline{s}_1 \wedge \underline{s}_2), \tag{4}$$

$$[\xi(\underline{s}_1), \eta(\underline{b}_2)] = -\eta(\underline{s}_1 \wedge \underline{b}_2), \tag{5}$$

$$[\eta(\underline{b}_1), \eta(\underline{b}_2)] = -k\xi(\underline{b}_1 \wedge \underline{b}_2). \tag{6}$$

Journal of Physics: Conference Series **545** (2014) 012001 doi:10.1088/1742-6596/545/1/012001

In particular, when $\underline{s}$ or $\underline{b}$ is equal to one of the basis vectors $\underline{e}_1 = (1,0,0)$, $\underline{e}_2 = (0,1,0)$ or $\underline{e}_3 = (0,0,1)$ it is sometimes convenient to introduce the infinitesimal generators $\xi_j := \xi(\underline{e}_j)$, $\eta_j := \eta(\underline{e}_j)$, $j = 1,2,3$ which are given explicitly by[1]

$$\xi_1 = y\frac{\partial}{\partial z} - z\frac{\partial}{\partial y} = -\sin\varphi\frac{\partial}{\partial\vartheta} - \cot\vartheta\cos\varphi\frac{\partial}{\partial\varphi}, \tag{7}$$

$$\xi_2 = z\frac{\partial}{\partial x} - x\frac{\partial}{\partial z} = \cos\varphi\frac{\partial}{\partial\vartheta} - \cot\vartheta\sin\varphi\frac{\partial}{\partial\varphi}, \tag{8}$$

$$\xi_3 = x\frac{\partial}{\partial y} - y\frac{\partial}{\partial x} = \frac{\partial}{\partial\varphi}, \tag{9}$$

and

$$\begin{aligned}\eta_1 &= \left(1+\frac{k}{4}x^2-\frac{k}{4}y^2-\frac{k}{4}z^2\right)\frac{\partial}{\partial x} + \frac{k}{2}xy\frac{\partial}{\partial y} + \frac{k}{2}xz\frac{\partial}{\partial z}\\ &= \left(1+\frac{k}{4}r^2\right)\sin\vartheta\cos\varphi\frac{\partial}{\partial r} + \frac{1}{r}\left(1-\frac{k}{4}r^2\right)\left(\cos\vartheta\cos\varphi\frac{\partial}{\partial\vartheta} - \frac{\sin\varphi}{\sin\vartheta}\frac{\partial}{\partial\varphi}\right),\end{aligned} \tag{10}$$

$$\begin{aligned}\eta_2 &= \frac{k}{2}xy\frac{\partial}{\partial x} + \left(1-\frac{k}{4}x^2+\frac{k}{4}y^2-\frac{k}{4}z^2\right)\frac{\partial}{\partial y} + \frac{k}{2}yz\frac{\partial}{\partial z}\\ &= \left(1+\frac{k}{4}r^2\right)\sin\vartheta\sin\varphi\frac{\partial}{\partial r} + \frac{1}{r}\left(1-\frac{k}{4}r^2\right)\left(\cos\vartheta\sin\varphi\frac{\partial}{\partial\vartheta} + \frac{\cos\varphi}{\sin\vartheta}\frac{\partial}{\partial\varphi}\right),\end{aligned} \tag{11}$$

$$\begin{aligned}\eta_3 &= \frac{k}{2}xz\frac{\partial}{\partial x} + \frac{k}{2}yz\frac{\partial}{\partial y} + \left(1-\frac{k}{4}x^2-\frac{k}{4}y^2+\frac{k}{4}z^2\right)\frac{\partial}{\partial z}\\ &= \left(1+\frac{k}{4}r^2\right)\cos\vartheta\frac{\partial}{\partial r} - \frac{1}{r}\left(1-\frac{k}{4}r^2\right)\sin\vartheta\frac{\partial}{\partial\vartheta},\end{aligned} \tag{12}$$

where we have also given the expressions in terms of spherical coordinates (r,ϑ,φ) such that $\underline{x} = (x,y,z) = r(\sin\vartheta\cos\varphi, \sin\vartheta\sin\varphi, \cos\vartheta)$. The commutation relations (4,5,6) imply:

$$[\xi_1,\xi_2] = -\xi_3, \qquad [\xi_1,\eta_2] = -\eta_3, \qquad [\eta_1,\eta_2] = -k\xi_3, \tag{13}$$

and cyclic permutations. Note that for the spatially flat case, $k=0$, ξ_j and η_j are the infinitesimal generators of rotations and translations, respectively. When $k \neq 0$ the space Σ_k is still spherically symmetric, so that the generators ξ_j are the same as for $k=0$, but in this case the η_j's cannot be interpreted as generators of translations.

2.2. Most general homogeneous and isotropic distribution function on a FRW spacetime

In order to derive the most general distribution function on a FRW spacetime which is spatially homogeneous and isotropic, we compute the lifted vector fields $\hat{\xi}(\underline{s})$ and $\hat{\eta}(\underline{b})$. In terms of adapted local coordinates (x^μ, p^μ) on the tangent bundle TM, the lift $\hat{\xi}$ of a Killing vector field ξ is given by (see, for instance [15])

$$\hat{\xi} = \xi^\mu\frac{\partial}{\partial x^\mu} + p^\alpha\frac{\partial\xi^\mu}{\partial x^\alpha}\frac{\partial}{\partial p^\mu}. \tag{14}$$

[1] The Killing vector fields η_a correspond to the vector fields $\mathbf{X}_a$, $a = 1,2,3$, in Ref. [18], and the vector fields ξ_a to the vector fields $\mathbf{X}_{3+a}$, $a = 1,2,3$, in that reference. However, notice that while the expressions in terms of Cartesian coordinates agree precisely with those of Eq. (2) in [18], it seems there are misprints in the expressions for $\mathbf{X}_3$ and $\mathbf{X}_4$ in spherical coordinates in Eq. (3) of that reference and in Eq. (2) of Ref. [19]: the expression for $\mathbf{X}_4$ is off by a minus factor and the last term in the expression for $\mathbf{X}_3$ should have a derivative with respect to θ instead of ϕ.

For the lift of the Killing vector fields defined in Eqs. (2,3) this yields the following expressions:

$$\begin{aligned}\hat{\xi}(\underline{s}) &= (\underline{s}\wedge\underline{x})\cdot\frac{\partial}{\partial\underline{x}} + (\underline{s}\wedge\underline{p})\cdot\frac{\partial}{\partial\underline{p}},\\ \hat{\eta}(\underline{b}) &= \left(1-\frac{k}{4}|\underline{x}|^2\right)\underline{b}\cdot\frac{\partial}{\partial\underline{x}} + \frac{k}{2}\left[(\underline{b}\cdot\underline{x})\left(\underline{x}\cdot\frac{\partial}{\partial\underline{x}}+\underline{p}\cdot\frac{\partial}{\partial\underline{p}}\right)+(\underline{b}\cdot\underline{p})\underline{x}\cdot\frac{\partial}{\partial\underline{p}}-(\underline{x}\cdot\underline{p})\underline{b}\cdot\frac{\partial}{\partial\underline{p}}\right],\end{aligned}$$

where here and below, $\underline{p} := (p^x, p^y, p^z)$.

Now consider a smooth function $f : TM \to \mathbb{R}$ on the tangent bundle TM which is G-symmetric, that is, invariant with respect to the flows generated by $\hat{\xi}(\underline{s})$ and $\hat{\eta}(\underline{b})$ for all $\underline{s},\underline{b}\in\mathbb{R}^3$, and let us characterize such functions. We start with the simpler case $k=0$ of a spatially flat universe. As commented above, in this case the Killing vector fields $\eta(\underline{b})$ generate translations in directions of $\underline{b}$, and we have simply

$$\hat{\eta}(\underline{b}) = \underline{b}\cdot\frac{\partial}{\partial\underline{x}}.$$

Therefore, invariance with respect to these translations implies that the distribution function must have the form

$$f = f(\tau, p^\tau, p^x, p^y, p^z), \tag{15}$$

that is, independent of $\underline{x}$. Then, invariance with respect to rotations implies that

$$(\underline{s}\wedge\underline{p})\cdot\frac{\partial f}{\partial\underline{p}} = 0,$$

for all $\underline{s}\in\mathbb{R}^3$ which, in turn, implies that

$$f(x,p) = F(\tau, p^\tau, |\underline{p}|),\qquad |\underline{p}| := \sqrt{(p^x)^2+(p^y)^2+(p^z)^2},$$

as can be shown by introducing spherical coordinates for $\underline{p}$, for instance.

When $k\neq 0$ the vector fields $\hat{\eta}(\underline{b})$ on the tangent bundle TM are more complicated. However, it is not difficult to deduce the form of the invariant distribution function after the following two observations. First, it is convenient to replace the adapted local coordinates (x^μ, p^μ) used so far with new local coordinates $(x^\mu, p^{a'})$ on TM, where $p^{a'}$ refer to the components of p with respect to a local orthonormal frame. Specifically, we define

$$p^{0'} := p^\tau,\qquad \underline{p}' := \frac{a(\tau)}{1+\frac{k}{4}|\underline{x}|^2}\underline{p},$$

such that $g(p,p) = -(p^{0'})^2 + |\underline{p}'|^2$. With respect to these new coordinates the lifted generators assume the slightly simpler form

$$\hat{\xi}(\underline{s}) = (\underline{s}\wedge\underline{x})\cdot\frac{\partial}{\partial\underline{x}} + (\underline{s}\wedge\underline{p}')\cdot\frac{\partial}{\partial\underline{p}'}, \tag{16}$$

$$\hat{\eta}(\underline{b}) = \left(1-\frac{k}{4}|\underline{x}|^2\right)\underline{b}\cdot\frac{\partial}{\partial\underline{x}} + \frac{k}{2}\left[(\underline{b}\cdot\underline{x})\underline{x}\cdot\frac{\partial}{\partial\underline{x}}+(\underline{b}\cdot\underline{p}')\underline{x}\cdot\frac{\partial}{\partial\underline{p}'}-(\underline{x}\cdot\underline{p}')\underline{b}\cdot\frac{\partial}{\partial\underline{p}'}\right]. \tag{17}$$

The second observation is based on the vector identity

$$(\underline{b}\cdot\underline{p}')\underline{x} - (\underline{x}\cdot\underline{p}')\underline{b} = (\underline{b}\wedge\underline{x})\wedge\underline{p}',$$

which allows us to rewrite $\hat{\eta}(\underline{b})$ in the simpler form

$$\hat{\eta}(\underline{b}) = \left(1 + \frac{k}{4}|\underline{x}|^2\right)\underline{b}\cdot\frac{\partial}{\partial\underline{x}} + \frac{k}{2}\hat{\xi}(\underline{b}\wedge\underline{x}). \tag{18}$$

Therefore, if f is invariant with respect to the flows generated by $\hat{\xi}(\underline{s})$ and $\hat{\eta}(\underline{s})$, it follows that $\pounds_{\hat{\xi}(\underline{b}\wedge\underline{x})}f = 0$ and we conclude that f must have the same form as in Eq. (15) except that $\underline{p} = (p^x, p^y, p^z)$ should be replaced by $\underline{p}'$. Then, $\pounds_{\hat{\xi}(\underline{s})}f = 0$ implies as before that f can only depend on τ, $p^{0'}$ and $|\underline{p}'|$.

We conclude that the most general spatially homogeneous and isotropic distribution function on a FRW spacetime has the form

$$f(x,p) = F\left(\tau, p^\tau, \frac{|\underline{p}|}{1+\frac{k}{4}|\underline{x}|^2}\right), \qquad |\underline{p}| := \sqrt{(p^x)^2+(p^y)^2+(p^z)^2}, \tag{19}$$

for some function F. If we only consider a simple gas of particles with the same positive rest mass $m > 0$, then it is sufficient to work on the associated future mass shell Γ_m^+, from which

$$p^\tau = \sqrt{m^2 + a(\tau)^2\frac{|\underline{p}|^2}{\left(1+\frac{k}{4}|\underline{x}|^2\right)^2}}$$

can be eliminated.

2.3. Most general collisionless, homogeneous and isotropic distribution function

Our approach so far was merely based on symmetry considerations. Now we wish to derive the most general distribution function f on a FRW spacetime which, in addition of being spatially homogeneous and isotropic, is also collisionless. This means that f has to satisfy the Liouville (or collisionless Boltzmann) equation

$$\pounds_L f = p^\mu\frac{\partial f}{\partial x^\mu} - \Gamma^\mu{}_{\alpha\beta}p^\alpha p^\beta\frac{\partial f}{\partial p^\mu} = 0. \tag{20}$$

For the derivation of f, we first recall the following Lemma (see, for instance, Proposition 6 in Ref. [15]):

Lemma 1. *Let ξ be the generator of a one-parameter group of isometries on (M,g). Then, the quantity $Q_\xi(x,p) := g_x(p,\xi)$ is conserved along the flow generated by the Liouville vector field L.*

As a consequence of this lemma, the six quantities $J_a := Q_{\xi_a} = g(p,\xi_a)$, $K_a := Q_{\eta_a} = g(p,\eta_a)$, $a = 1,2,3$, are conserved along the flow of L. For the following, we consider the important quantity

$$C^2 := |\underline{K}|^2 + k|\underline{J}|^2 = \sum_{a=1,2,3}\left(K_a^2 + kJ_a^2\right), \tag{21}$$

which is also conserved along the Liouville flow. As the next lemma shows, it follows from the commutation relations (4,5,6) that C^2 is also invariant with respect to the flows generated by the vector fields $\hat{\xi}(\underline{s})$ and $\hat{\eta}(\underline{b})$.

Lemma 2. *We have $\pounds_{\hat{\xi}(\underline{s})}C^2 = \pounds_{\hat{\eta}(\underline{b})}C^2 = 0$ for all $\underline{s},\underline{b}\in\mathbb{R}^3$.*

Proof. In order to show this we first note the following identity which is valid for any two Killing vector fields ξ, η on (M, g):

$$\pounds_{\hat{\xi}} Q_\eta = Q_{[\xi,\eta]}, \tag{22}$$

where $Q_\eta(x,p) = g_x(p,\eta)$ is the conserved quantity associated to η. This identity follows almost immediately from the fact that the lifted generator $\hat{\xi}$ is a Killing vector field of $(TM, \hat{g})$ that commutes with the Liouville vector field L, see section 5 in Ref. [15]. Using these observations we find

$$\pounds_{\hat{\xi}} Q_\eta = \pounds_{\hat{\xi}}[\hat{g}(\hat{\eta}, L)] = \hat{g}(\pounds_{\hat{\xi}}\hat{\eta}, L) = \hat{g}([\hat{\xi},\hat{\eta}], L) = g([\xi,\eta],p) = Q_{[\xi,\eta]},$$

where we have also used Lemma 10 in Ref. [15] in the last step.

Using the identity (22) and the commutation relations (13) it follows for all $i = 1, 2, 3$:

$$\begin{aligned}
\pounds_{\hat{\xi}_i} C^2 &= 2\sum_{a=1,2,3} \left(K_a Q_{[\xi_i,\eta_a]} + kJ_a Q_{[\xi_i,\xi_a]}\right) = -2\sum_{a,b=1,2,3} \varepsilon_{iab}(K_a K_b + kJ_a J_b) = 0,\\
\pounds_{\hat{\eta}_i} C^2 &= 2\sum_{a=1,2,3} \left(K_a Q_{[\eta_i,\eta_a]} + kJ_a Q_{[\eta_i,\xi_a]}\right) = -2\sum_{a,b=1,2,3} \varepsilon_{iab}(kK_a J_b + kJ_a K_b) = 0,
\end{aligned}$$

which proves the lemma. □

Therefore, the quantity C^2 is invariant with respect to both the Liouville flow and the symmetry group. An explicit calculation based on the expressions (2,3) reveals that

$$C = a(\tau)^2 \frac{|\underline{p}|}{1 + \frac{k}{4}|\underline{x}|^2}, \qquad |\underline{p}| := \sqrt{(p^x)^2 + (p^y)^2 + (p^z)^2}. \tag{23}$$

Comparing this with Eq. (19) we see that the most general spatially homogeneous and isotropic distribution function on a FRW spacetime can also be written as

$$f(x,p) = F(\tau, m, C), \tag{24}$$

for some smooth function F. The previous Lemma provides an alternative explanation for the invariance of this f under the symmetry group G: it is a function of the three quantities τ, m and C, all of which are invariant with respect to the flows generated by the lifted generators of the symmetry group. The arguments presented in the previous subsection show that this is the most general distribution function which is G-symmetric.

Introducing Eq. (24) into the Liouville equation, it is straightforward to derive from this the most general G-symmetric, collisionless distribution function: Since m and C are conserved along the Liouville flow, the Liouville equation (20) reduces simply to

$$p^\tau \frac{\partial F}{\partial \tau} = 0,$$

implying that F is independent of τ. Therefore, we arrive at the simple result that the most general *collisionless*, spatially homogeneous and isotropic distribution function on a FRW spacetime must have the form

$$f(x,p) = F(m, C). \tag{25}$$

3. Observables: current density and stress-energy tensor

In this section, we evaluate the most important observables of the theory, namely the current density J and the stress-energy tensor T assuming a distribution function of the form $f(x,p) = F(m,C)$, see Eq. (25). We only consider a simple gas, where each gas particle has the same rest mass $m > 0$. Therefore, since m is fixed, we omit the explicit dependance of F from

Journal of Physics: Conference Series **545** (2014) 012001 doi:10.1088/1742-6596/545/1/012001

m in what follows. The observables J and T are defined as fiber integrals over the future mass hyperboloidal

$$P_x^+ := \{p \in T_xM : g_x(p,p) = -m^2, p \text{ future directed}\}$$

with associated volume element π_x, and their geometrical invariant definition is given by (see [14, 15])

$$J_x(\alpha) := \int_{P_x^+} f(x,p)\alpha(p)\pi_x \qquad \text{(current density)}, \tag{26}$$

$$T_x(\alpha,\beta) := \int_{P_x^+} f(x,p)\alpha(p)\beta(p)\pi_x \qquad \text{(stress-energy tensor)}, \tag{27}$$

where $\alpha, \beta \in T_x^*M$ are covectors at $x \in M$. Since J_x is linear in α and T_x is bilinear in (α, β), they define a vector field and a symmetric, contravariant tensor field on M, respectively. Their components relative to local coordinates (x^μ) of M are obtained from $J^\mu(x) = J_x(dx^\mu)$, $T^{\mu\nu}(x) = T_x(dx^\mu, dx^\nu)$, which yields

$$J^\mu(x) = \int_{P_x^+} f(x,p)p^\mu \pi_x, \tag{28}$$

$$T^{\mu\nu}(x) = \int_{P_x^+} f(x,p)p^\mu p^\nu \pi_x. \tag{29}$$

As long as f satisfies the Liouville equation, these observables are divergence-free, and if f is nonnegative it follows that J is future directed while T satisfies the familiar energy conditions (for a proof of these assertions see, for instance, Ref. [14, 15]).

For a spatially homogeneous and isotropic distribution function $f(x,p) = F(C)$ the right-hand sides of Eqs. (28,29) can be simplified. For this we introduce an orthonormal basis in the tangent space T_xM, defined by

$$(e_0, e_x, e_y, e_z) := \left(\frac{\partial}{\partial\tau}, \frac{1+\frac{k}{4}|\underline{x}|^2}{a(\tau)}\frac{\partial}{\partial x}, \frac{1+\frac{k}{4}|\underline{x}|^2}{a(\tau)}\frac{\partial}{\partial y}, \frac{1+\frac{k}{4}|\underline{x}|^2}{a(\tau)}\frac{\partial}{\partial z} \right), \tag{30}$$

so that any vector $p \in P_x^+$ can be written in the form

$$p = p^\mu \frac{\partial}{\partial x^\mu} = p^{0'} e_0 + p^{x'} e_x + p^{y'} e_y + p^{z'} e_z$$

with $p^{0'} = \sqrt{m^2 + (p^{x'})^2 + (p^{y'})^2 + (p^{z'})^2}$. Passing to a spherical polar representation, we get the equivalent form

$$p = \sqrt{m^2 + |\underline{p}'|^2} e_0 + |\underline{p}'| n^i e_i, \tag{31}$$

with $(n^x, n^y, n^z) = (\sin\theta\cos\phi, \sin\theta\sin\phi, \cos\theta)$. Relative to these frames, the Lorentz-invariant volume element π_x takes the form

$$\pi_x = \frac{dp^{x'} \wedge dp^{y'} \wedge dp^{z'}}{\sqrt{m^2 + (p^{x'})^2 + (p^{y'})^2 + (p^{z'})^2}} = \frac{|\underline{p}'|^2 d|\underline{p}'| \wedge (\sin\theta d\theta) \wedge d\phi}{\sqrt{m^2 + |\underline{p}'|^2}}, \tag{32}$$

while for a spatially homogeneous and isotropic distribution function belonging to a simple gas we have

$$f(x,p) = F(C) = F(a(\tau)|\underline{p}'|), \tag{33}$$

where the quantity C has been defined in Eq. (23), and we recall that here $|\underline{p}'|$ refers to the norm of the spatial components of $p \in P_x^+$ with respect to the *orthonormal* basis defined in Eq. (30), such that $C = a(\tau)|\underline{p}'|$.

Substituting Eqs. (32,33) into Eq. (28), respectively Eq. (29), we find that the frame components of the current density and the stress-energy tensor take the form

$$J^{\mu'}(x) = \int_0^\infty\int_0^\pi\int_0^{2\pi} F(a(\tau)|\underline{p}'|)p^{\mu'}\frac{|\underline{p}'|^2\sin\theta d\phi d\theta d|\underline{p}'|}{\sqrt{m^2+|\underline{p}'|^2}}, \tag{34}$$

$$T^{\mu'\nu'}(x) = \int_0^\infty\int_0^\pi\int_0^{2\pi} F(a(\tau)|\underline{p}'|)p^{\mu'}p^{\nu'}\frac{|\underline{p}'|^2\sin\theta d\phi d\theta d|\underline{p}'|}{\sqrt{m^2+|\underline{p}'|^2}}. \tag{35}$$

Using Eq. (31) and changing the variable of integration from $|\underline{p}'|$ to $C = a(\tau)|\underline{p}'|$ we find that the current density is described by

$$J(x) = J^{\mu'}(x)e_{\mu'} = \left(\frac{4\pi}{a^3(\tau)}\int_0^\infty F(C)C^2dC\right)e_0, \tag{36}$$

i.e. J is parallel to the four-velocity of the isotropic observers (those that see the universe homogenous and isotropic). This property of J originates in the spatially homogenous and isotropic property of the distribution function. A similar computation shows that the nonvanishing frame components of the stress-energy tensor take the form:

$$T^{0'0'} = \frac{4\pi}{a^4(\tau)}\int_0^\infty F(C)\sqrt{(ma(\tau))^2+C^2}C^2dC, \tag{37}$$

$$T^{x'x'} = T^{y'y'} = T^{z'z'} = \frac{4\pi}{3a^4(\tau)}\int_0^\infty F(C)\frac{C^4}{\sqrt{(ma(\tau))^2+C^2}}dC. \tag{38}$$

Therefore, the current density and stress-energy tensor have the same form as the one of a perfect fluid, for which

$$J^\mu = n(\tau)u^\mu, \qquad T^{\mu\nu} = [\rho(\tau)+P(\tau)]u^\mu u^\nu + P(\tau)g^{\mu\nu}$$

with $u = \frac{\partial}{\partial\tau}$ the dynamical mean velocity of the gas, $n(\tau)$ the particle density and $\rho(\tau)$ the energy density of the gas measured by observers comoving with the mean velocity, and $P(\tau)$ is the mean or isotropic pressure measured by such observers. In our case, the densities and pressure are given by

$$n(\tau) = \frac{4\pi}{a^3(\tau)}\int_0^\infty F(C)C^2dC, \tag{39}$$

$$\rho(\tau) = \frac{4\pi}{a^4(\tau)}\int_0^\infty F(C)\sqrt{(ma(\tau))^2+C^2}C^2dC, \tag{40}$$

$$P(\tau) = \frac{4\pi}{3a^4(\tau)}\int_0^\infty F(C)\frac{C^4}{\sqrt{(ma(\tau))^2+C^2}}dC. \tag{41}$$

Once the function $F(C)$ and the scale factor $a(\tau)$ have been specified, the current density and the stress-energy tensor of the gas are uniquely determined.

The expressions in Eqs. (40,41) imply

$$\rho(\tau) = 3P(\tau) + \frac{4\pi m^2}{a^2(\tau)} \int_0^\infty F(C) \frac{C^2}{\sqrt{(ma(\tau))^2 + C^2}} dC,$$

and thus we always have $\rho \geq 3P \geq 0$ since the distribution function is nonnegative. These inequalities imply that the stress-energy tensor satisfies the weak, the strong and the dominant energy conditions, as is generally true for a kinetic gas (cf. Ref. [14]). It is also a simple matter to verify the equations $\nabla_\mu J^\mu = 0$, $\nabla_\mu T^{\mu\nu} = 0$, which in the spatially homogeneous and isotropic case reduce to

$$\frac{dn}{d\tau} + 3n\frac{d\log a}{d\tau} = 0, \qquad \frac{d\rho}{d\tau} + 3(\rho + P)\frac{d\log a}{d\tau} = 0, \tag{42}$$

from the explicit expressions given in Eqs. (39,40,41).

Finally, we observe that in the limit where $ma \to 0$ which occurs either as $m \to 0$ (photon gas), or for fixed $m > 0$ as $a \to 0$, one gets the equation of state $P = \frac{1}{3}\rho$. Therefore, in these limits the gas behaves as a radiation fluid. Furthermore, expanding the integrals for the case in which the scale factor approaches zero it is seen that the energy density and pressure both diverge as a^{-4}, as for a radiation fluid. In the opposite limit, that is, when a becomes very large, the energy density and pressure decay to zero as a^{-3} and a^{-5}, respectively, and the collisionless kinetic gas behaves more and more like dust. In order to see this more explicitly, it is instructive to consider the particular case for which the distribution function is supported in a very narrow interval around some constant $C = C_0 > 0$:

$$F(C) = \frac{N}{4\pi C^2}\delta(C - C_0), \tag{43}$$

for some dimensionless constant $N > 0$, where δ denotes the Dirac distribution. In this case, we obtain from Eqs. (39,40,41),

$$n(\tau) = \frac{N}{a^3(\tau)}, \qquad \rho(\tau) = \frac{N}{a^4(\tau)}\sqrt{(ma(\tau))^2 + C_0^2}, \qquad P(\tau) = \frac{N}{3a^4(\tau)}\frac{C_0^2}{\sqrt{(ma(\tau))^2 + C_0^2}}, \tag{44}$$

and we see that in the limit $ma(\tau) \ll C_0$, $\rho(\tau) \simeq 3P(\tau) \simeq NC_0/a^4(\tau)$, while in the limit $C_0 \to 0$ or $ma(\tau) \gg C_0$, $\rho(\tau) \simeq Nm/a^3(\tau)$ and $P(\tau) \simeq 0$.

In the next section, we combine the results obtained here with Einstein's field equations and discuss the evolution of a universe which is filled with a simple, collisionless gas, with or without a cosmological constant.

4. The evolution of the universe

In the previous sections we have derived the one-particle distribution function f and the associated current density and stress-energy tensor for a spatially homogeneous and isotropic, simple, collisionless gas propagating on a fixed FRW background spacetime. The distribution function is characterized by a function $F(C)$ of the single variable C defined in Eq. (23), and once the scale factor $a(\tau)$ is determined, the stress-energy tensor is known. In order to determine $a(\tau)$ in a self-consistent way, we have to solve Einstein's field equations. This is the subject of the present section.

4.1. Formulation through a one-dimensional mechanical system

For the FRW metric (1) Einstein's field equations reduce to (see, for instance [26]) the Friedmann equations

$$\frac{3}{a^2}\left(\frac{da}{d\tau}\right)^2 = 8\pi G\rho + \Lambda - \frac{3k}{a^2}, \tag{45}$$

$$\frac{3}{a}\frac{d^2a}{d\tau^2} = -4\pi G(\rho + 3P) + \Lambda, \tag{46}$$

with Newton's constant G and the cosmological constant Λ. Here, we assume that the energy density ρ and isotropic pressure P are those of a simple, collisionless gas, see Eqs. (40,41), and hence P and ρ depend themselves on the scale factor $a(\tau)$. Since by construction the expressions for ρ and P given in Eqs. (40,41) satisfy Eq. (42) and since Eq. (46) is a consequence of Eq. (45) and Eq. (42), it is sufficient to consider the first Friedmann equation (45) in what follows. Once the function $F(C)$ has been chosen, this equation yields a first-order differential equation for the scale factor $a(\tau)$.

The following analysis is based on the simple observation that the first Friedmann equation (45), together with the expression for ρ given in Eq. (40), describes a mechanical system for the motion of a one-dimensional particle in a potential. In order to describe this system it is convenient to introduce the following dimensionless quantities:

$$x(t) := \frac{a(\tau)}{a_0}, \qquad t := H_0\tau, \qquad f(\zeta) := (ma_0)^3 F(C), \qquad \zeta := C/(ma_0),$$

where $a_0 = a(\tau = 0)$ is the present value of the scale factor and $H_0 := d\log a(\tau)/d\tau|_{\tau=0}$ is the Hubble constant. Note that by definition, $x(0) = \dot{x}(0) = 1$, where here and in the following the dot denotes differentiation with respect to the dimensionless time parameter t. In terms of these quantities Eq. (45) can equivalently be written as

$$\dot{x}^2 + V(x) = \Omega_c, \tag{47}$$

with the effective potential

$$V(x) = -\frac{\mu}{x^2}\int_0^\infty f(\zeta)\sqrt{x^2+\zeta^2}\zeta^2 d\zeta - \Omega_\Lambda x^2, \tag{48}$$

where we have defined $\mu := 32\pi^2 Gm/(3a_0^3H_0^2)$ and introduced the usual density parameters $\Omega_\Lambda := \Lambda/(3H_0^2)$ and $\Omega_c := -k/(H_0^2a_0^2)$ (see, for instance [26]). Evaluating Eq. (47) at $t = 0$ yields the familiar constraint

$$1 = \Omega_M + \Omega_\Lambda + \Omega_c, \qquad \Omega_M := \mu\int_0^\infty f(\zeta)\sqrt{1+\zeta^2}\zeta^2 d\zeta$$

for the density parameters.

Therefore, the dynamics of the scale factor $a(\tau)$ determining the evolution of the universe is equivalent to the dynamics of a one-dimensional point particle with energy Ω_c propagating in the potential $V(x)$ given in Eq. (48) whose initial conditions are $x(0) = \dot{x}(0) = 1$. Consequently, it is sufficient to analyze the basic properties of the potential $V(x)$ in order to understand the qualitative features of the evolution of the universe.

As long as f is nonnegative and decays sufficiently fast such that Ω_M is finite, the effective potential $V(x)$ is well-defined for all $x > 0$. Furthermore, using Lebesgue's dominated convergence theorem, it is not difficult to check that

$$\lim_{x\to 0} x^2 V(x) = -\mu \int_0^\infty f(\zeta)\zeta^3 d\zeta,$$

while

$$\lim_{x\to\infty} x(V(x) + \Omega_\Lambda x^2) = -\mu \int_0^\infty f(\zeta)\zeta^2 d\zeta.$$

By successive differentiation of $V(x)$ we obtain

$$\frac{dV(x)}{dx} = \frac{\mu}{x^3}\int_0^\infty f(\zeta)\frac{x^2+2\zeta^2}{\sqrt{x^2+\zeta^2}}\zeta^2 d\zeta - 2\Omega_\Lambda x, \tag{49}$$

$$\frac{d^2V(x)}{dx^2} = -\frac{\mu}{x^4}\int_0^\infty f(\zeta)\frac{2x^4+9x^2\zeta^2+6\zeta^4}{(x^2+\zeta^2)^{3/2}}\zeta^2 d\zeta - 2\Omega_\Lambda. \tag{50}$$

4.2. The dynamics for $\Lambda = 0$

When the cosmological constant is zero ($\Omega_\Lambda = 0$), it follows from Eqs. (48,49,50) that $V(x)$ is a negative, monotonously increasing and concave function with asymptotics $V(x) \simeq -C_3/x^2$ for $x \to 0$ and $V(x) \simeq -C_2/x$ for $x \to \infty$ where C_2, C_3 are positive constants. Therefore, the potential looks qualitatively like the function plotted in dotted red in Fig. 1.

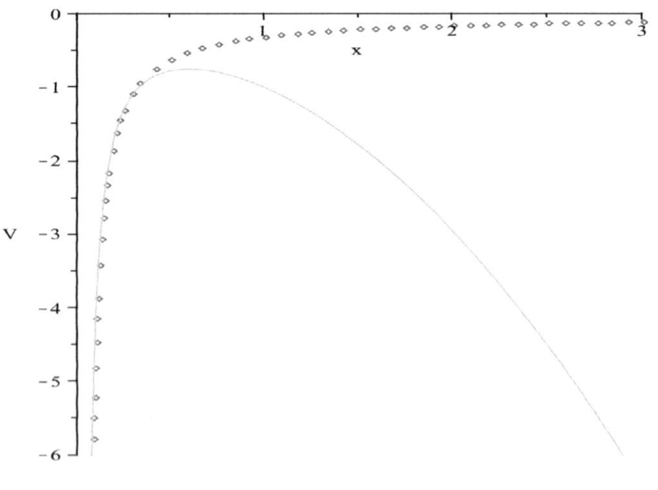

Figure 1. The effective potential $V(x)$ for the case where the distribution function has the form given in Eq. (43) with $C_0/(ma_0) = 0.5$. The parameters used are $\Omega_M = 0.3$, $\Omega_\Lambda = 0.0$ (red dotted line) and $\Omega_M = 0.3$, $\Omega_\Lambda = 0.7$ (green solid line).

From this we can immediately reach the following conclusions: For spatially flat ($k = 0$) or hyperboloid ($k = -1$) universes, the trajectory $x(t)$ exists for all times in the interval

$t \in (-t_s, \infty)$, where

$$t_s = \int_0^1 \frac{dx}{\sqrt{\Omega_c - V(x)}} < \infty \tag{51}$$

is the age of the universe in units of Hubble time H_0^{-1}. The dimensionless scale factor $x(t)$ increases monotonically from $x = 0$ to $x = \infty$, describing a universe which continuously expands from the big bang singularity. The expansion is decelerating since $-dV/dx(x) < 0$. As $\tau \to \infty$, $a(\tau)$ grows like $\tau^{2/3}$ and τ, respectively, depending on whether $k = 0$ or $k = -1$, and ρ decays as $1/a^3(\tau)$ and P as $1/a^5(\tau)$ as can be inferred from Eqs. (40,41).

For a universe with compact spherical sections ($k = 1$) the energy of the particle $\Omega_c < 0$ is negative, implying a turning point in the particle's trajectory. Therefore, in this case the universe expands from the big bang singularity, reaches a maximal volume and recollapses to a crunch singularity in a finite amount of proper time.

At the big bang or crunch singularities the scale factor $a(\tau)$ goes to zero and both the energy density $\rho(\tau)$ and the isotropic pressure $P(\tau)$ diverge like $1/a^4(\tau)$ as can be deduced from Eqs. (40,41).

4.3. The dynamics for $\Lambda > 0$

Recent observations suggest that the universe is currently undergoing an accelerated stage of expansion. However, as we have seen above, an accelerated expansion cannot be explained by a collisionless kinetic gas with nonnegative distribution function, since the effective potential $V(x)$ is monotonously increasing in this case. In fact, this can also be seen directly from the second Friedmann equation (46) which shows that any matter field with positive $\rho + 3P$ leads to deceleration.

The simplest manner to explain the current observations is to introduce a cosmological constant Λ which is large enough such that $\Lambda > 4\pi G(\rho + 3P)$ for late enough times, inducing an accelerated expansion. For this reason, in what follows we analyze the behavior of the effective potential $V(x)$ given in Eq. (48) assuming $\Omega_\Lambda > 0$.

When $\Omega_\Lambda > 0$, it still follows from Eqs. (48,50) that $V(x)$ is negative and concave, and that $V(x) \simeq -C_3/x^2$ for $x \to 0$. However, due to the presence of the cosmological constant, the potential now diverges for $x \to \infty$ since $V(x) \simeq -\Omega_\Lambda x^2$ in this limit. There is a unique maximum $x_* > 0$ of $V(x)$ which, according to Eq. (49), is determined by the implicit equation

$$2\Omega_\Lambda x_*^4 = \mu \int_0^\infty f(\zeta) \frac{x_*^2 + 2\zeta^2}{\sqrt{x_*^2 + \zeta^2}} \zeta^2 d\zeta. \tag{52}$$

Therefore, the effective potential looks qualitatively like the function plotted in solid green in Fig. 1. For a spatially flat ($k = 0$) or hyperboloid ($k = -1$) universe, for which $\Omega_c \geq 0$, the trajectory $x(t)$ exists again for all times $t \in (-t_s, \infty)$, with t_s given by Eq. (51), and the universe continuously expands from the big bang singularity. However, now the expansion is only decelerating until the time $t = t_*$ the maximum of the potential is reached: $x(t_*) = x_*$. After that time, the effects of the cosmological constant become important and the universe starts its phase of accelerated expansion. At late stages the scale factor $a(\tau)$ grows exponentially fast in τ.

For a universe with compact spherical sections ($k = 1$), $\Omega_c < 0$ and the qualitative behavior depends on whether or not Ω_c is larger than the maximum value of V. If $\Omega_c > V(x_*)$, the qualitative behavoir of $x(t)$ is the same as in the previous cases $k = 0, -1$. However, when $\Omega_c < V(x_*)$, there are two trajectories; one describing a universe that expands from the big bang, reaches a maximal volume and recollapses to a crunch singularity in a finite amount of proper time and one that contracts from $x = \infty$ to a minimal value and starts expanding again to $x = \infty$. If we accept the fact that there was a big bang in the past, the first trajectory is relevant and $x_* > x(0) = 1$. Combining Eqs. (48,52) with each other we obtain the estimate

$$-3\Omega_\Lambda x_*^2 < V(x_*) < -2\Omega_\Lambda x_*^2$$

for the maximal value of V, and consequently $\Omega_c < V(x_*)$ and $x_* > 1$ can only occur if

$$\Omega_c < -2\Omega_\Lambda.$$

Recent cosmological observations indicate that the current values of $(\Omega_M, \Omega_\Lambda, \Omega_c)$ lie around $(0.3, 0.7, 0.0)$, implying that this inequality cannot hold. Furthermore, by evaluating the slope of the potential today we obtain from Eq. (49),

$$\frac{dV(1)}{dx} = \mu \int_0^\infty f(\zeta) \frac{1+2\zeta^2}{\sqrt{1+\zeta^2}} \zeta^2 d\zeta - 2\Omega_\Lambda \leq 2(\Omega_M - \Omega_\Lambda),$$

which is negative, indicating that we are currently in the phase of accelerated expansion.

5. Conclusions

In this work, we have discussed a few basic properties of the Einstein-Liouville system under the assumption that the spacetime metric and the distribution function which describes a collisionless, uncharged, simple, massive gas is homogenous and isotropic. We have defined in detail this homogeneity and isotropicity property of the distribution function, and we have shown that these symmetry requirements and the Liouville equation restrict considerably its dependance upon the phase space variables. This simplification allows one to cast the Einstein-Liouville system (with or without cosmological constant) into a single, ordinary differential equation for the scale factor which can be interpreted as a one-dimensional mechanical system with an effective potential. The analysis of this system shows that irrespectively of the curvature of the spatial sections, any spatially homogeneous and isotropic universe has a singular origin where the curvature, the density and pressure of the gas become unbounded. Close to the singularity, the collisionless gas behaves as a radiation fluid. Furthermore, spatially flat and hyperboloidal universes expand forever, and as we have shown, at late times the kinetic gas behaves much like a pressureless perfect fluid. In this context, we also mention the recent work by Ringström [6] which proves the nonlinear stability of the spatially flat model with positive cosmological constant in the expanding direction.

It ought to be mentioned that the conclusion of the present work should be viewed with caution. The collisionless nature of the gas may be incompatible with the unbounded growth of the density. One expects that at high densities the mean free path between collisions becomes very small, suggesting that one may not be allowed to neglect collisions between the gas particles. We will return to this question in future work, where the relativistic Boltzmann equation for a homogeneous and isotropic kinetic gas on a FRW background will be analyzed.

Acknowledgments

This work was supported in part by a CIC Grant from Universidad Michoacana.

References

[1] Bancel D and Choquet-Bruhat Y 1973 *Comm. Math. Phys.* **33** 83–96

[2] Rendall A 2004 *The Einstein equations and the large scale behavior of gravitational fields* ed Chrusciel P and Friedrich H (Basel: Birkhäuser) pp 231–250

[3] Rein G and Rendall A 1992 *Comm. Math. Phys.* **150** 561–583
[4] Dafermos M and Rendall A 2007 (*Preprint* gr-qc/0701034)
[5] Noutchegueme N and Tetsadjio M 2009 *Class. Quant. Grav.* **26** 195001
[6] H Ringström 2013 *On the Topology and Future Stability of the Universe* (Oxford: Oxford University Press)
[7] Andréasson H 2011 *Living Reviews in Relativity* **14** URL http://www.livingreviews.org/lrr-2011-4
[8] Synge J 1934 *Trans. Royal Soc. Canada* **28** 127–171
[9] Tauber G and Weinberg J 1961 *Phys. Rev.* **122** 1342–1365
[10] Israel W 1963 *J. Math. Phys.* **4** 1163–1181
[11] Lindquist R 1966 *Ann. Phys. (N.Y.)* **37** 487–518
[12] Ehlers J 1971 *General Relativity and Cosmology* ed Sachs R (New York: Acedemic) pp 1–70
[13] Ehlers J 1973 *Relativity, Astrophysics and Cosmology* ed Israel W (Dordrecht: D. Reidel) pp 1–125
[14] Sarbach O and Zannias T 2013 *AIP Conf. Proc.* **1548** 134–155
[15] Sarbach O and Zannias T 2014 *Class. Quantum Grav.* **31** 085013
[16] Sarbach O and Zannias T 2014 *AIP Conf. Proc.* **1577** 192–207
[17] Maartens R and Maharaj S 1987 *Gen. Rel. Grav.* **19** 499–509
[18] Maartens R and Maharaj S 1987 *Gen. Rel. Grav.* **19** 1217–1222
[19] Maartens R and Maharaj S 1987 *Gen. Rel. Grav.* **19** 1223–1234
[20] Ehlers J, Geren P and Sachs R 1968 *J. Math. Phys.* **9** 1344–1349
[21] Bel L 1969 *Astrophys. J.* **155** 83–87
[22] Alvarez E, Bel L and Gracia-Bondía J 1975 *Astron. and Astrophys.* **40** 381–386
[23] Ray J and Zimmerman J 1977 *Nuevo Cimento* **42B** 183–197
[24] Riess A and et al 1998 *Astronomical Journal* **116** 1009–1038
[25] Perlmutter S and et al 1999 *Astrophysical Journal* **517** 565–586
[26] Carroll S 2004 *Spacetime and Geometry: An introduction to General Relativity* (San Francisco, USA: Addison Wesley)

Journal of Physics: Conference Series **545** (2014) 012002 doi:10.1088/1742-6596/545/1/012002

Search for vacuum solutions of conformally flat stationary noncircular spacetime

Cuauhtemoc Campuzano[1,2] **and José Luis Méndez** [3]

[1]Instituto de Física, Pontificia Universidad Católica de Valparaíso, Casilla 4059, Valparaíso, Chile.

[2]Departamento de Física, Facultad de Física e Inteligencia Artificial, Universidad Veracruzana, 91000, Xalapa Veracruz, Mexico.

[3]Instituto de Física y Matemáticas, Universidad Michoacana de San Nicolás de Hidalgo. Edicio C-3, Cd. Universitaria, 58040 Morelia, Michoacán, México.

E-mail: ccampuzano@uv.mx

Abstract.
The main goal of this work is to present the first general vacuum solutions, for a conformally flat stationary and non-circular spacetime. The solutions found are static and have an axis of symmetry. In addition, we discuss the geometrical consequences related to the non-circular property of the solutions.

1. Introduction

The non-circular stationary spacetimes were first pointed out by Gourgoulhon and Bonazzola in the context of neutron stars with a strong toroidal magnetic field where such field is a consequence of the associated non-circular stress-energy tensor [1]. In this regards, Ioka and Sasaki introduced a general formalism based in the magnetohydrodynamics to study the general case [2], and then analyzed relativistic stars possessing both, poloidal and toroidal magnetic fields as well as a meridional flow [3]. Recently, attention has been paid to the study of neutron stars with meridional circulations [4]. In the exact solutions framework, the most general form of the conformally flat stationary and non-circular spacetimes was developed in [5]. This subject has been extensively studied by several authors and great part of the geometrical properties are now well understood [6, 7, 8].

On the theoretical side, the exact solutions comprise an active line of research in General Relativity where it was argued that the physically interesting spacetimes belong to the family of Lewis-Papapetrou metrics (L-P).The L-P spacetimes are characterized by the presence of two commuting Killing vectors one $\mathbf{k}$, timelike and other $\mathbf{m}$, spacelike; thus spacetime becomes to cyclic stationary. The cyclic simmetry is the most general which is related with the spacelike Killing vector, however, there is a symmetry related with both Killing vectors which isn't highlighted, we refer to the circular symmetry. Specifically, a stationary and cyclically symmetric spacetime is circular if the 2-surfaces orthogonal to the Killing fields are integrable. This is equivalent to satisfy the Frobenius integrability conditions or, alternatively, that the gravitational field is not only locally independent of the time and the rotation angle, but also invariant under the simultaneous inversion of the time and the angle. The circularity property

Journal of Physics: Conference Series **545** (2014) 012002 doi:10.1088/1742-6596/545/1/012002

imposes geometrical restrictions on the structure of the spacetime metrics, as a consequence the Killing vectors are restricted to be on a plane expanded by them. In the case where the circularity is not compulsory, the metric functions do not depend on the coordinates of the Killing vectors, and therefore the spacetime has the most general structure and the metric functions depend only of the two coordinates. These spacetimes are stationary non-circular and cyclic-symmetric.

In the Section 2 some geometrical facts and the vacuum solutions for the conformally flat stationary and non-circular spacetimes are presented, finally in Section 3 we present some conclusion and discussions .

2. Non-circular spacetimes

Usually, when one refers to a stationary axisymmetric spacetime, one immediately associates a Lewis-Papapetrou type metric to it. However, as it is remarked by some authors (see [9] and reference there in) in many works, the axial symmetry is not inherent to the general structure of a L-P metric. As it is well known this is an arrangement of fixed points which is invariant under the action of SO(3) group transformations. The circular symmetry is a property not emphasized enough like axial symmetry, and both are the restricted cases of non-circular and cyclic symmetries, respectively [10].

In this work we consider the search of vacuum solutions of non-circular stationary spacetimes; in particular, for the Petrov type 0 or conformally flat which was reported in [5] and is described by the metric,

$$ds^2 = e^{-2Q}\Big[- k(d\tau^2 + 2x d\tau d\sigma + \epsilon d\sigma^2) + 2k(K_0 + K_1 x) d\sigma dy + \frac{dx^2}{(C_0 + C_1 x)(x^2 - \epsilon) - k(K_0 + K_1 x)^2} + (C_0 + C_1 x) dy^2\Big], \tag{1}$$

where $Q = Q(x, y)$ and K_0, K_1, C_0 y C_1 are the integration constants. The parameters are $k = \pm 1$ and $\epsilon = \pm 1$, and basically its values define the temporal and angular coordinates in [9] a detailed study is given. In what follows and without lose of generality we use $k = 1$.

The spacetime described by the metric (1) has two commuting Killing vectors, one $\mathbf{k}$ timelike and other $\mathbf{m}$ spacelike and as a results of the non-circularity, are establishes

$$\begin{aligned}\star(\mathbf{m} \wedge \mathbf{k} \wedge d\mathbf{k}) &= e^{2Q}(K_0 + K_1 x),\\ \star(\mathbf{k} \wedge \mathbf{m} \wedge d\mathbf{m}) &= e^{2Q}(\epsilon K_1 + K_0 x),\end{aligned} \tag{2}$$

where the star stands for the Hodge dual. When the right hand side of the above equation vanish the spacetime is circular and (2) they are know as the circular conditions. So, when $K_0 = 0$ and $K_1 = 0$ we arrive to circular stationary spacetime and cyclic reported in [9]. Furthermore, contrary to what happens in the circular case, the non-circular has an axis of symmetry.

The main goal of this paper is to show exact solutions for the metric (1) in the case of vacuum [11][12]. The starting point are the Einstein field equations,

$$G_{\mu\nu} = 0 \qquad \Leftrightarrow \qquad R_{\mu\nu} = 0. \tag{3}$$

In order to solve the system of equations thrown by (3), we proceed to integrate the more easy equations, in our case from G^x_τ and G^x_σ we get $Q(x, y) = q(x)$ the next step is to integrate G^τ_σ

$$G^\tau_\sigma = e^{2Q}\Big\{[\epsilon(C_0 + C_1 x) + K_0(K_0 + K_1 x)]\frac{\partial Q}{\partial x} - \frac{(K_0 K_1 + C_1 \epsilon)}{2}\Big\} = 0,$$

integrate directly the above equation, we obtain

$$q(x) = \frac{1}{2}\ln[(K_0K_1 + \epsilon C_1)x + K_0^2 + \epsilon C_0] + \frac{1}{2}q_0, \tag{4}$$

where q_0 is an integration constant.

Substituting (4) into the equations system (3), now from the components ${}^y{}_\tau$ and ${}^y{}_\sigma$ we get

$$e^{q_0}\epsilon(C_0K_1 - K_0C_1) = 0, \tag{5}$$
$$e^{q_0}(K_0^3 - \epsilon K_1^2K_0 - \epsilon^2K_1C_1 + \epsilon K_0C_0) = 0, \tag{6}$$

from (5) one finds

$$K_1 = \frac{C_1}{C_0}K_0, \tag{7}$$

which substituting (7) in (6) one gets

$$\left(\epsilon\frac{C_1^2}{C_0^2} - 1\right)(K_0^2 + \epsilon C_0) = 0, \tag{8}$$

to satisfy the above equation we obtain the vacuum solution for conformally flat stationary and non-circular spacetime. It is remarkable at a glance for to $\epsilon = -1$ one can satisfy for the second term, however, in a deep analysis the second factor should not vanish in order to have a well defined metric.

2.1. Vacuum solution

When $\epsilon = 1$ from the equation (8) seemingly we have two possibilities, but actually it is just one, namely

$$i)\left(\frac{C_1^2}{C_0^2} - 1\right) = 0 \Rightarrow |C_1| = |C_0|, \tag{9}$$

and then C's, to have the same sign or contrary sign.

For the first case of (i), when $C_1 = C_0$ the Eq.(7) implies $K_1 = K_0$, then we found

$$i.a)\quad q(x) = \frac{1}{2}\ln[q_0(K_0^2 + C_0)(1 + x)]. \tag{10}$$

In the second case, with contrary signs, says $C_1 = -C_0$ implies $K_1 = -K_0$ and then

$$i.b)\quad q(x) = \frac{1}{2}\ln[q_0(K_0^2 + C_0)(1 - x)]. \tag{11}$$

In summary, we have $\epsilon = 1$, $C_0 = C_1$, $K_0 = K_1$ and the conformal factor of the metric (1) for the case ($i.a$) is read as

$$e^{-2Q} = \frac{1}{q_0(K_0^2 + C_0)(1 + x)}. \tag{12}$$

For the case $(i.b)$ $\epsilon = 1$, $C_0 = -C_1$, $K_0 = -K_1$ and now we have the conformal factor given by

$$e^{-2Q} = \frac{1}{q_0(K_0^2 + C_0)(1 - x)}. \tag{13}$$

In [5] it was shown that (1) is static if the metric functions satisfy $\epsilon = 1$, and $K_0 = \pm K_1$, and a consequence of the integration of the field equations $C_0 = \pm C_1$, which means that this branch of solutions is static. Another fact is the existence of an axis of symmetry at $x = \mp 1$.

In general for $k = 1$, $\epsilon = 1$, the metric of the spacetime is rewritten as

$$ds^2 = \frac{1}{q_0(K_0^2 + C_0)(1 \pm x)}\Big[- dt^2 + (x^2 - 1)d\sigma^2 + 2K_0(x \pm 1)d\sigma dy$$
$$+ \frac{dx^2}{C_0(x \pm 1)(x^2 - \epsilon) - K_0(x \pm 1)^2} + C_0(x \pm 1)dy^2\Big], \tag{14}$$

where $dt \to d\tau + x d\sigma$. The metric (14) shows the static structure explicitly and the axis of symmetry.

3. Conclusions

In this work we wish to emphasize the non-circular property in the stationary spacetimes, which was unknown for several years, and recently brought to scene. We present general solutions to the Einstein's equations in vacuum for a non-circular stationary spacetime and conformally flat. The solution is found to be static and axisymmetric and it shows an unusual structure because the metric is not diagonal. Unfortunately the Riemann tensor vanishes for the solutions given, which means that are flat spacetimes.

The solutions were obtained by solving a couple of linear equations given by Einstein equations, the integral gives the function of scale factor (4), which after of substituting in the others equations we have the constraints (5) and (6). Like any system of equations these must be satisfied simultaneously, so therefore, our result is just one of the possibilities still lacking to investigate the remaining linearly independent solutions of the field equations, but this will be done in another job.

Similarly to what happens in the circular stationary spacetimes there is a vacuum solution for the static branch and it has an axis of symmetry, as demonstrated in [5]. In fact, the noncircular stationary metrics have not been explored in depth because the technical difficulties, but we find quite interesting that the development of such solutions could be related with several astrophysical phenomena. We think that there are more interesting consequences which can only be studied in non-circular geometries whose study will be done elsewhere.

Acknowledgments

CC acknowledges partial support by CONACyT Grant CB-2012-177519-F, CONACyT Grant I0010-2014-02 Estancias Internacionales-233618-C and grant PROMEP, CA-UV, Álgebra, Geometría and Gravitación. This work was partially supported by SNI (México). CC specially thank to Efraín Rojas and Gerardo Torres del Castillo for enlightening discussions.

References

[1] Gourgoulhon E and Bonazzola S 1993 *Phys. Rev. D* **48**, 2635.
[2] Ioka K and Sasaki M 2003 *Phys. Rev. D* **67**, 12026, arXiv:gr-qc/0302106.
[3] Ioka K and Sasaki M 2004 *Astrophys.J.* **600**, 296-316, arXiv:astro-ph/0305352.
[4] Birkl R, Stergioulas N and Muller E 2011 *Phys. Rev. D* **84**, 023003, arXiv:gr-qc/1011.5475.

[5] Ayón-Beato E, Campuzano C and García A 2006 *Phys. Rev. D***74**, 024014, arXiv:gr-qc/0506004.
[6] Vera R 2003 *Class. Quant. Grav.* **20**, 2785, arXiv:gr-qc/0305108.
[7] Anabalon A, Deruelle N, Morisawa Y, Oliva J, Sasaki M, Tempo D and Troncoso R 2009 *Class. Quant. Grav.* **26**, 065002, arXiv:0812.3194[hep-th].
[8] Anabalon A, Tempo D and Troncoso R 2011 *Int. J. Mod. Phys.* **20**, 639-647, arXiv:gr-qc/1009.3030.
[9] García A and Campuzano C 2002 *Phys. Rev. D***66**, 124018 (2002), arXiv:gr-qc/0205080. [Erratum-ibid 2003 D **68**, 049901 (2003)].
[10] García A and Campuzano C 2003 *Circular stationary cyclic symmetric space-times: Conformal flatness* arXiv:gr-qc/0310054.
[11] Méndez Montes de Oca, José Luis, *Tesis: El problema de Kerr y los espacios estacionarios no-circulares*, Departamento de Física, Facultad de Física e I. A., Universidad Veracruzana 2011.
[12] Campuzano Vargas, Cuauhtemoc, *Tesis: Solución de vacío para espacios estacionarios con dos simetrías y conformalmente planos*, Facultad de Ciencias Físico Matemáticas, Benemérita Universidad Autónoma de Puebla 2014.

Journal of Physics: Conference Series **545** (2014) 012003 doi:10.1088/1742-6596/545/1/012003

Vacuum force for a massive scalar field in a multiply warped braneworld

V. E. Ceron

E-mail: vceron@uaeh.edu.mx
Área Académica de Matemáticas y Física, Universidad Autónoma del Estado de Hidalgo, Carretera Pachuca-Tulancingo Km. 4.5, C P. 42184, Pachuca, México.

Héctor H. Hernández

E-mail: hhernandez@uach.mx
Universidad Autónoma de Chíhuahua, Facultad de Ingeniería, Nuevo Campus Universitario, Chihuahua, Chih., México.

L. A. López

E-mail: lalopez@uaeh.edu.mx
Área Académica de Matemáticas y Física, Universidad Autónoma del Estado de Hidalgo, Carretera Pachuca-Tulancingo Km. 4.5, C P. 42184, Pachuca, México.

Omar Pedraza

E-mail: omarp@uaeh.edu.mx
Área Académica de Matemáticas y Física, Universidad Autónoma del Estado de Hidalgo, Carretera Pachuca-Tulancingo Km. 4.5, C P. 42184, Pachuca, México.

Abstract. We determine the finite temperature Casimir force for a scalar massive field for a pair of parallel plates in a 6D non-factorizable geometry. In this contribution we use the Green's function method to determine the Casimir force. It can be expressed in terms of two four-dimensional Casimir force contributions: one for the zero mode and the other for a tower of massive modes associated with the extra dimensions. An uncertainty band of 10% around the theoretical standard 4D Casimir force was set up, by limiting the effective 4D Casimir force within such band yields the bound $k/r_z \gg 1.97 \times 10^6$ TeV.

1. Introduction

The idea of considering our observable four-dimensional universe as a subspace of a higher dimensional spacetime is seriously considered in high energy physics. These models with extra dimensions were proposed to solve various fundamental problems such as the problem of hierarchy among others (see [1, 2, 3]). Several ideas in this direction have evolved to explore various implications in cosmology, astrophysics and particle physics.

A recent work [4] has considered extension of the Randall-Sundrum model to more than one extra dimension, where the warped compact dimensions get successive more

Journal of Physics: Conference Series **545** (2014) 012003 doi:10.1088/1742-6596/545/1/012003

warping. The resulting geometry of the D dimensional spacetime reduces to $M^{1,D-1} \rightarrow \{[M^{1,3} \times S^1/Z_2] \times S^1/Z_2\} \times \ldots$ with $D-4$ warped spacelike dimensions. One of the interesting characteristics of such a model is that the resulting geometry is similar to a combination RS and ADD scheme of compactification. The Kaluza-Klein (KK) modes for a massive scalar field and gravitational field have been studied in [5] and [6].

The prevalence of spacetimes with extra dimensions has motivated research on possible corrections to Casimir effect acting between parallel plates due to the existence of extra dimensions. In general the Casimir effect may be defined as the stress on the boundary surface when a quantum field in its vacuum state is confined to a finite volume of space.

In this work we analyse the Casimir force in this model as a test for the phenomenological viability of such multiply warped spacetime. To obtain the Casimir force for the 6D massive scalar field we use the Green's function method.

2. The model

We work in a 6D non-factorizable geometry, which is doubly warped and compactified with a Z_2 orbifold in each of the extra dimensions [4]. The metric g_{MN} in this scenario is given by

$$ds_6^2 = b^2(z)[a^2(y)\eta_{\mu\nu}dx^\mu dx^\nu - R_y{}^2dy^2] - r_z{}^2dz^2\,, \tag{1}$$

greek indexes denote 4D space-time coordinates, $\eta_{\mu\nu}$ is the Minkowski metric with signature $(+,-,-,-)$, while the coordinates y and z represent the extra spatial dimensions. R_y and r_z are the two moduli corresponding to the y and z directions and $a(y)$ and $b(z)$ are the corresponding warp factors. The warp factors are given by

$$a(y) = e^{-\rho|y|}\,, \qquad \rho = \frac{R_y k}{r_z \cosh k\pi}\,, \tag{2}$$

$$b(z) = \frac{\cosh(kz)}{\cosh(k\pi)}\,, \qquad k = r_z\sqrt{\frac{-\Lambda}{M^4}}\,, \tag{3}$$

where M is the 6D Planck constant and Λ the bulk cosmological constant.

It is worth mentioning that this model can be seen as a box-like picture of the bulk, where the walls of the box are 4-branes, and the intersection of two 4-branes may be identified with a 3-brane. The Standard Model brane (SM brane) depends on the values of the parameters, which are determined in terms of Λ, r_z, R_y. We can identify our SM brane by requiring the desired TeV scale, therefore we must identify the SM brane with the one at $y = \pi$, $z = 0$, while the 3-brane located at $y = 0$, $z = \pi$ must be identified with the Planck brane. The other two 3-branes located at $y = 0$, $z = 0$, and $y = \pi$, $z = \pi$ have intermediate energy scale.

For the resolution of the hierarchy problem in this model we need that

$$10^{-16} = \frac{e^{-\rho\pi}}{\cosh k\pi}\,. \tag{4}$$

This equation, along with the condition (2), implies large warping in one direction and small in the other.

Finally, we mention that in this model the KK modes of bulk scalar fields have been studied [5], however, when the higher dimensional scalar field is massive, there does not exist a zero mode solution, so in order to overcome this problem we would now address the nature of Kaluza Klein modes of scalar fields in the six dimensional braneworld.

3. Massive scalar field

In this section we start by computing the mass spectrum and eigenfunctions for the massive scalar field. Let us consider the 6D action for a massless scalar field Φ in the metric (1)

$$S = \frac{1}{2}\int d^4x \int_{-\pi}^{\pi} dy \int_{-\pi}^{\pi} dz\, \sqrt{-g}\, \left[g^{MN}\partial_M\Phi\partial_N\Phi - M_\Phi^2\Phi^2\right], \tag{5}$$

where g_{MN} is the six dimensional metric (1). As it is well known [7], when the higher dimensional scalar field is massive there does not exist a zero mode solution. In order to obtain a zero mode we need to modify the action an include boundary mass terms, as it was made in [8]. Thus we propose the following boundary mass terms

$$S' = -\int d^4x \int_{-\pi}^{\pi} dy \int_{-\pi}^{\pi} dz\, \sqrt{-g}\left\{\frac{2\tilde{c}\rho}{R^2 b^2(z)}\left[\delta(y) - \delta(y-\pi)\right] + \frac{2\tilde{d}k}{r_z^2}\left[\delta(z) - \delta(z-\pi)\right]\right\}\Phi^2, \tag{6}$$

where $\tilde{c}$ and $\tilde{d}$ are dimensionless constants. Assuming solutions of the form $\Phi(x^\mu, y, z) = \phi(x^\mu)\,\xi(y)\chi(z)$ and performing separation of variables in the field equations for the massive scalar field, it is straightforward to obtain the three differential equations

$$\begin{aligned} \left(\Box_4 + m_{nN}^2\right)\phi(x^\mu) &= 0, && (7)\\ \frac{1}{R_y^2}\partial_y\left(a^4\partial_y\xi(y)\right) + m_n^2 a^4\xi(y) - m_{nN}^2 a^2\xi(y) - \frac{2\tilde{c}a^4}{R^2}\rho\left[\delta(y) - \delta(y-\pi)\right]\xi(y) &= 0, && (8)\\ \frac{1}{r_z^2}\partial_z\left(b^5\partial_z\chi(z)\right) + m_n^2 b^3\chi(z) - M^2 b^5\chi(z) - \frac{2\tilde{d}k}{r_z^2}b^5\left[\delta(z) - \delta(z-\pi)\right]\chi(z) &= 0. && (9) \end{aligned}$$

Notice that m_n and m_{nN} correspond to the spectra of the modes for the two compact coordinates z and y respectively. From the 4D point of view Eq. (7) corresponds to an effective massive scalar field $\phi(x^\mu)$ whose mass m_{nN}, picks up two independent contribution corresponding to the y and z coordinates. Our first task is to determine this mass spectrum and the associated eigenfunctions.

3.1. Mode descomposition

In this subsection we focus on the y, z dependence, Eqs. (8) and (9). To obtain the KK mass spectrum we first determine the tower denoted by m_n and then using it as the input to the Eq. (8) to determine the mass tower m_{nN}.

In order to implement the boundary conditions in the z direction, we integrate the equation (9) over an interval across the two boundaries at $z = 0, \pi$, respectively, which allows us to match the modes across the brane along the coordinate z.

For reasonable values of the parameter k, we consider the following form of the warp factor $b(z) \sim e^{-k(\pi - z)} = e^{-kz'}$. Performing the change of variable $\tilde{z} = \frac{r_z m_n}{k}e^{kz'}$ and redefining the function as $\tilde{\chi} = e^{-\nu kz'}\chi$, $\nu = 5/2$, one gets the Bessel equation;

$$\tilde{z}^2\frac{d^2\tilde{\chi}_n}{d\tilde{z}^2} + \tilde{z}\frac{d\tilde{\chi}_n}{d\tilde{z}} + (\tilde{z}^2 - \alpha^2)\tilde{\chi}_n = 0, \tag{10}$$

where $\alpha = \sqrt{\nu^2 + \frac{M^2 r_z^2}{k^2}}$. When $m_{n=0} = 0$ the solution has the form

$$\chi_0(z) = \begin{cases} \sqrt{\frac{(3+2\tilde{d})k}{2r_z\left(e^{(3+2\tilde{d})k\pi}-1\right)}}, & \frac{3}{2}+\tilde{d} > 0 \text{ localized in } z=\pi \\ \frac{1}{\sqrt{2r_z\pi}}, & \frac{3}{2}+\tilde{d} = 0 \text{ no localization} \\ \sqrt{\frac{(2\tilde{d}+3)k}{2r_z\left(1-e^{(2\tilde{d}+3)k\pi}\right)}}, & \frac{3}{2}+\tilde{d} < 0 \text{ localized in } z=0 \end{cases}, \qquad \int_{-\pi}^{\pi} dz\, r_z b^3(z)\chi_0^2(z) = 1 \tag{11}$$

In the case $m_n > 0$ the normalized modes become

$$\chi_n(z) = \frac{e^{\nu k z'}}{d_1}\left[J_\alpha\left(\frac{r_z m_n}{k} e^{kz'}\right) + d_2 Y_\alpha\left(\frac{r_z m_n}{k} e^{kz'}\right)\right], \tag{12}$$

where $d_{1,2}$ are arbitrary constants and J_α and Y_α are Bessel's functions of order α. The constants in (12) are chosen in such a way that the orthogonality relations are

$$\int_{-\pi}^{\pi} dz\, b^3(z)\chi_n(z)\chi_{n'}(z) = \delta_{nn'}, \tag{13}$$

fulfilling the boundary condition $\left[\frac{d\chi_n(z)}{dz} - \tilde{d}k\chi_n(z)\right]\Big|_{z=0,\pi} = 0$. The KK masses m_n are determined by imposing the boundary condition on the solution (12). Notice that the appropriate boundary condition depends on whether the 5D field is odd or even under the orbifold Z_2 symmetry. In this work we consider that the 5D field is even under the Z_2 symmetry. Thus we have

$$\chi_n(z) \to \chi_n(-z). \tag{14}$$

Imposing the boundary conditions in the low energy regime $m_n r_z \ll k$ and $kr_z \gg 1$, we have the approximated KK mode masses

$$m_n \simeq \left(n + \frac{\alpha}{2} - \frac{3}{4}\right)\frac{\pi k e^{-k\pi}}{r_z}, \qquad n = 1, 2, \dots, \qquad \alpha > 0. \tag{15}$$

After obtaining the z dependent part of the KK modes and the corresponding spectrum, we now can solve the y dependent part of the modes. Adopting a similar technique, we can rewrite the Eq. (8) as

$$\tilde{y}^2\frac{d^2\tilde{\xi}_N}{d\tilde{y}^2} + \tilde{y}\frac{d\tilde{\xi}_N}{d\tilde{y}} + (\tilde{y}^2 - \beta^2)\tilde{\xi}_N = 0, \tag{16}$$

here $\beta = \sqrt{4 + \left(\frac{m_n R}{\rho}\right)^2}$ and, once again,we have performed the change of variable $\tilde{y} = R m_{nN} e^{\rho y}/\rho$ and redefining the function as $\xi = \tilde{y}^2\tilde{\xi}$. It is worth mentioning that the equation (8) is invariant under a change of sign in the coordinate y and therefore it is enough to solve the equation in the region $y > 0$.

When $m_{n=0,N=0} = 0$, the solution of Eq. (16) yields a zero mode

$$\xi_0(y) = \begin{cases} \sqrt{\frac{(\tilde{c}-1)\rho}{R\left(e^{(\tilde{c}-1)\rho\pi}-1\right)}}, & \tilde{c}-1 > 0 \text{ localized in } y=\pi \\ \frac{1}{\sqrt{2R\pi}}, & \tilde{c}-1 = 0 \text{ no localization}, \\ \sqrt{\frac{(\tilde{c}-1)\rho}{R\left(1-e^{(\tilde{c}-1)\rho\pi}\right)}}, & \tilde{c}-1 < 0 \text{ localized in } y=0 \end{cases}, \qquad \int_{-\pi}^{\pi} dy\, a^2(y)\xi_0(y)\xi_0(y) = 1. \tag{17}$$

Journal of Physics: Conference Series **545** (2014) 012003 doi:10.1088/1742-6596/545/1/012003

As for the massive modes, we have

$$\xi_N(y) = \frac{e^{2\rho|y|}}{c_1}\left[J_\beta\left(\frac{m_{nN}R}{k}e^{\rho|y|}\right) + c_2 Y_\beta\left(\frac{m_{nN}R}{k}e^{\rho|y|}\right)\right], \tag{18}$$

where $c_{1,2}$ are arbitrary constants, and J_β, Y_β are Bessel's functions of order β. $\xi_N(y)$ satisfies the normalization and boundary conditions

$$\int_{-\pi}^{\pi} dy\, Ra^2(y)\xi_N(y)\xi_{N'}(y) = \delta_{NN'}, \tag{19}$$

$$\left[\frac{d\xi(y)}{dy} - \tilde{c}\rho\xi(y)\right]\Bigg|_{y=0,\pi} = 0. \tag{20}$$

The KK masses are determined by imposing the boundary conditions (20) and, in the limit $m_{nN}R \ll \rho$ and $\rho R \gg 1$, allows us to obtain the approximated KK tower

$$m_{0N} \simeq \left[N - \frac{1}{2}\right]\frac{\rho e^{-\rho\pi}}{R}, \quad N = 1, 2, \ldots, m_{n=0} = 0, \tag{21}$$

$$m_{nN} \simeq \left[N + \sqrt{\frac{1}{2} + \left[\left(n + \sqrt{\frac{25}{16} + \left(\frac{M_\Phi r_z}{2k}\right)^2} - \frac{3}{4}\right)\frac{kRe^{-k\pi}}{2r_z\rho}\right]^2 - \frac{3}{4}}\right]\frac{\rho e^{-\rho\pi}}{R},$$
$$N = 1, 2, \ldots, m_n \neq 0. \tag{22}$$

We have obtained the mass spectrum of the scalar field modes by looking at the y, z dependence, our next task is to solve Eq. (7) for the x dependence and incorporate them all to get the Green's function.

3.2. Green's function

As a first step in the calculation of the Casimir force one has to find the Green's function for the 6D field equations. The corresponding Green's function satisfies

$$\Bigg[\frac{1}{b^2a^2}\Box_4 - \frac{1}{\sqrt{-g}}\partial_y\left(\sqrt{-g}\partial_y\right) - \frac{1}{\sqrt{-g}}\partial_z\left(\sqrt{-g}\partial_z\right) + M^2 + \frac{2\tilde{c}\rho}{R^2b^2}\left[\delta(y) - \delta(y-\pi R)\right]$$
$$+\frac{2\tilde{d}k}{r_z^2}\left[\delta(z) - \delta(z - \pi r_z)\right]\Bigg] G_{6D}(x, y, z, x', y', z') = \frac{\delta(x-x')\delta(y-y')\delta(z-z')}{\sqrt{-g}}. \tag{23}$$

We solve this equation using the eigenfunction expansion for the Green's function

$$G_6(x, y, z, x', y', z') = \sum_n\sum_N \chi_n^*(z')\chi_n(z)\xi_N^*(y')\xi_N(y)G_4(x, x'), \tag{24}$$

with $*$ denoting complex conjugation and G_{4D} is the 4D Green's function. To determine the explicit form of the 4D Green's function we need to solve the equation (7). To obtain the solutions for ϕ it is convenient to split up the 3D position vector as follows, $\vec{x} = (\vec{x}_\perp, x_3)$, where x_3 is the coordinate orthogonal to the plates and $\vec{x}_\perp$ denotes a 2D vector in the direction of the unbounded space coordinates along the plates. For the geometry of parallel plates the 4D Green's function is given by (see for example [9])

$$G_{4D}(x, x') = \int \frac{d\omega}{2\pi}\frac{d^2k_\perp}{(2\pi)^2}e^{-i\omega(t-t') + i\vec{k}_{perp}\cdot(\vec{x}_{perp} - \vec{x}'_{perp})}g(x_3, x_3'), \tag{25}$$

where

$$g(x_3,x_3') = \begin{cases} -\frac{1}{\lambda\sin(\lambda\ell)}\sin(\lambda x_{3<})\sin[\lambda(x_{3>}-\ell)]\,, & 0\le x_3,x_3'\le\ell\,,\\ \frac{1}{\lambda}\sin[\lambda(x_{3<}-\ell)]e^{i\lambda(x_{3>}-\ell)}\,, & \ell\le x_3,x_3'\,,\end{cases} \tag{26}$$

here ℓ is the separation between plates, $\lambda^2=\omega^2-\vec{k}_\perp^2-m_{nN}^2$, $x_{3>}(x_{3<})$ represents the greater (lesser) of x_3 and x_3'.

4. Casimir Force

Knowing the Green's function we can calculate the force on the bounding surfaces from the stress tensor. This can be done evaluating the discontinuity in the flux of the stress tensor across the plate, i. e., the corresponding normal-normal component of the stress tensor at $x_3=\ell$

$$F=\int_0^A dx_\perp\int_{-\pi}^{\pi}dy\int_{-\pi}^{\pi}dz\,\sqrt{|g_{\text{plate}}|}\left[\langle T_{33}\rangle_{in}|_{x_3=\ell}-\langle T_{33}\rangle_{out}|_{x_3=\ell}\right]\,, \tag{27}$$

where A is the area of the plate and g_{plate} is the induced metric on the physical plate located a $x_3=\ell$, $\sqrt{|g_{\text{plate}}|}$ contributes with the term $Rr_za^2b^3$.

As it is well known [9], the normal-normal component of the stress tensor is related to the Green's function through

$$\langle T_{33}\rangle_{\text{in}}|_{x_3=\ell} \sim \frac{1}{2i}\partial_{x_3}\partial_{x_3'}G_{\text{in}}(x_3,x_3')\Big|_{x_3=x_3'=\ell}=\frac{i}{2}\lambda\cot(\lambda\ell)\,, \tag{28}$$

$$\langle T_{33}\rangle_{\text{out}}|_{x_3=\ell} \sim \frac{1}{2i}\partial_{x_3}\partial_{x_3'}G_{\text{out}}(x_3,x_3')\Big|_{x_3=x_3'=\ell}=\frac{\lambda}{2}\,. \tag{29}$$

So, now we can find the force per unit area on the conducting surface

$$f_{\text{eff}}\equiv\frac{F}{A}=\sum_n\sum_N\int\frac{d\omega d^2k_\perp}{(2\pi)^3}\left(\frac{i}{2}\lambda\cot(\lambda\ell)-\frac{\lambda}{2}\right)\,, \tag{30}$$

where the dependence on the y and z coordinates vanishes completely by relations (13) and (19). Using dimensional regularization (see [10] for more details) we obtain that

$$f_{\text{eff}}=-\sum_{n,N}2\left(\frac{m_{nN}}{4\pi}\right)^{\frac{d+2}{2}}\left[\frac{1}{\ell^{\frac{d+2}{2}}}\sum_{\kappa=1}^{\infty}\frac{K_{\frac{d+2}{2}}(2\kappa\ell m_{nN})}{\kappa^{\frac{d+2}{2}}}-\frac{2m_{nN}}{\ell^{\frac{d}{2}}}\sum_{\kappa=1}^{\infty}\frac{K_{\frac{d+4}{2}}(2\kappa\ell m_{nN})}{\kappa^{\frac{d}{2}}}\right]\,, \tag{31}$$

and we can rewrite the full Casimir force as

$$f_{\text{eff}}=-\sum_{n=0}^{\infty}\sum_{N=0}^{\infty}\frac{m_{nN}^2}{8\pi^2}\left[\frac{1}{\ell^2}\sum_{\kappa=1}^{\infty}\frac{K_2(2\kappa\ell m_{nN})}{\kappa^2}-\frac{2m_{nN}}{\ell}\sum_{\kappa=1}^{\infty}\frac{K_3(2\kappa\ell m_{nN})}{\kappa}\right]\,, \tag{32}$$

where we have taken the limit $d\to2$ in (31). Sometimes this expression is presented in a slightly different way, which can be obtained by using the identity $K_\nu(z)=K_{\nu-2}(z)+\frac{2(\nu-1)}{z}K_{\nu-1}(z)$. In this case

$$f_{\text{eff}}=-\sum_{n=0}^{\infty}\sum_{N=0}^{\infty}\frac{m_{nN}^2}{8\pi^2}\left[\frac{3}{\ell^2}\sum_{\kappa=1}^{\infty}\frac{K_2(2\kappa\ell m_{nN})}{\kappa^2}+\frac{2m_{nN}}{\ell}\sum_{\kappa=1}^{\infty}\frac{K_1(2\kappa\ell m_{nN})}{\kappa}\right]\,. \tag{33}$$

The above result (33) represents the Casimir force due to a massive scalar field Φ. It is worth mentioning that for $n=0$, N can have both zero and non-zero values. But for $n\neq0$, N can have only nonzero values, as is evident from equations (21)-(22). Thus, we can write the Casimir force as

$$
\begin{aligned}
f_{\text{eff}} &= -\frac{\pi^2}{480}\frac{1}{\ell^4} - \sum_{N=1}^{\infty}\frac{m_{0N}^2}{8\pi^2}\left[\frac{3}{\ell^2}\sum_{\kappa=1}^{\infty}\frac{K_2(2\kappa\ell m_{0N})}{\kappa^2} + \frac{2m_{0N}}{\ell}\sum_{\kappa=1}^{\infty}\frac{K_1(2\kappa\ell m_{0N})}{\kappa}\right] \\
&\quad - \sum_{n=1}^{\infty}\sum_{N=1}^{\infty}\frac{m_{nN}^2}{8\pi^2}\left[\frac{3}{\ell^2}\sum_{\kappa=1}^{\infty}\frac{K_2(2\kappa\ell m_{nN})}{\kappa^2} + \frac{2m_{nN}}{\ell}\sum_{\kappa=1}^{\infty}\frac{K_1(2\kappa\ell m_{nN})}{\kappa}\right] .
\end{aligned}
\tag{34}
$$

To complete the analytical study, we proceed now to evaluate numerically f_{eff}. Using the fact that $e^{\rho\pi} = 10^{16}/\cosh(k\pi)$ (as demanded by the hierarchy problem) and the definition of $\rho = Rk/[r_z \cosh(k\pi)]$, the equation (34) depends on two extra free parameters in addition to the plates separation ℓ: the mass of the 6-dimensional scalar field M_Φ and the size of one extra space dimension r_z (more precisely k/r_z).

In Fig. 1, the Casimir force as a function of the plate separation ℓ is shown for two k/r_z values. Due to the warp factor along of the y direction, the Casimir force is very sensitive to the k/r_z value. It should be noted that the measurements of Casimir force between parallel plates are in the 0.5 μm to 2 μm range [11].

Let us set an uncertainty 10% for the theoretical standard 4D Casimir force. This gives rise to an uncertainty band for the Casimir force for a massless scalar field. Thus, limiting f_{eff} within such band for different values of k/r_z ($k/r_z = M_\Phi$), we obtain the following bound for $k/r_z \gg 1.97 \times 10^6$ Tev (see the figure 1).

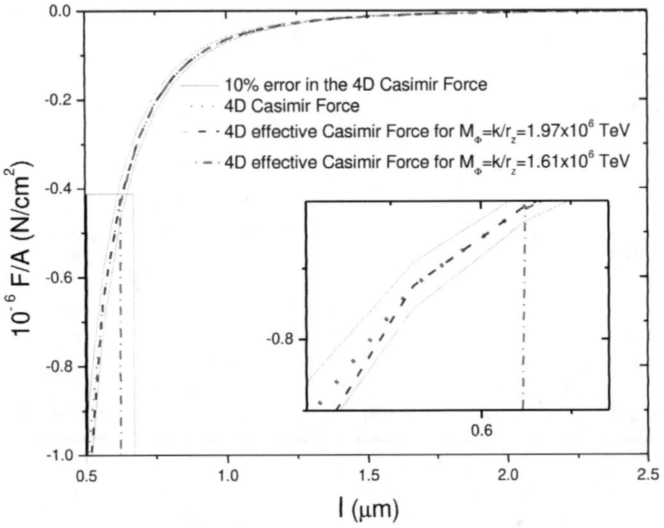

Figure 1. The Casimir force for a massive scalar field is shown as a function of the separation between plates for two values of k/r_z. In order to get a rough bound on k/r_z we also include a 10% error in the Casimir force. Such a region is represented by the two solid lines.

5. Discussion

In this work we have studied the Casimir force between two parallel plates in a doubly warped compactified six-dimensional spacetime with a Z_2 orbifolding in each of the extra dimensions.

Journal of Physics: Conference Series **545** (2014) 012003 doi:10.1088/1742-6596/545/1/012003

We adopt the Green's function approach to do so, therefore we first analized the modes for the 6D massive scalar field.

The KK modes can be obtained by imposing the boundary condition on the solutions for the scalar field, however, the appropiate boundary conditions depends on the parity of the scalar field under the Z_2 symmetry, thus without loss of generality we have choosen the fact that the 6D massive scalar field is even under the orbifold simmetry in each of the extra dimensions.

To satisfy the boundary condition at $z = 0$, $z = \pi$, $y = 0$ and $y = \pi$ the argument of the eigenfunctions (Bessel functions) has to satisfy a general equation which yields a discrete KK spectrum (22) in the limit $m_n r_z \ll k$, $kr_z \gg 1$ and (21) in the limit $m_{nN} R \ll k$, $\rho R \gg 1$.

Having summarized the KK tower for a massive scalar field, we proceeded with the calculation of the Casimir force. The mass of the KK modes yield a Casimir force naturally splitting into a leading order term given by a massless mode of the KK tower plus a correction term coming from the massive KK modes. The massive KK modes contain mixed information of the two extra dimensions (22). Usually in other works of the Casimir force in braneworlds the effective mass in 4D for a massles scalar field is the sum of the tower KK for each extra dimension [12, 13]. Such leading order term coincide with the standard 4D Casimir force for a massless scalar field.

We have also studied numerically the effective Casimir force (34) due to the extra dimensions. To do this, an uncertainty band around the theoretical force for a massless scalar field $f_{sf}^{4D} = -\hbar c\pi^2/(480\ell^4)$ was set, corresponding to an uncertainty of the 10%. Limiting $f_{\rm eff}$ within such band yields the bound for the scenario parameter.

In summary, it should be noted that although the Green's function technique involves further details of the mode decomposition of the corresponding fields, it is due to their orthogonality relations which involve the correct hypervolume factor of the plates, that one can eliminate the mode eigenfunctions from the Casimir force, one can conclude that localization of the field modes does not play a role as far as the Casimir force is concerned.

6. Acknowledgments

We thank Hugo A. Morales-Técotl for discussions and comments on this paper. HH was supported by Conacyt grant CB-2008-01/101774.

References

[1] Randall L and Sundrum R 1999 *Phys.Rev.Lett.* **83** 4690–4693 (*Preprint* hep-th/9906064)
[2] Randall L and Sundrum R 1999 *Phys.Rev.Lett.* **83** 3370–3373 (*Preprint* hep-ph/9905221)
[3] Arkani-Hamed N, Dimopoulos S and Dvali G 1998 *Phys.Lett.* **B429** 263–272 (*Preprint* hep-ph/9803315)
[4] Choudhury D and SenGupta S 2007 *Phys.Rev.* **D76** 064030 (*Preprint* hep-th/0612246)
[5] Koley R, Mitra J and SenGupta S 2010 *Europhys.Lett.* **91** 31001 (*Preprint* 1001.2666)
[6] Mukhopadhyaya B, Sen S and SenGupta S 2013 *J.Phys.* **G40** 015004 (*Preprint* 1106.1027)
[7] Gherghetta T and Pomarol A 2001 *Nucl.Phys.* **B602** 3–22 (*Preprint* hep-ph/0012378)
[8] Gherghetta T and Pomarol A 2000 *Nucl.Phys.* **B586** 141–162 (*Preprint* hep-ph/0003129)
[9] Milton K 2001
[10] Linares R, Morales-Tecotl H A and Pedraza O 2010 *Phys.Rev.* **D81** 126013 (*Preprint* 1003.4286)
[11] Frank M, Turan I and Ziegler L 2007 *Phys.Rev.* **D76** 015008 (*Preprint* 0704.3626)
[12] Frank M, Saad N and Turan I 2008 *Phys.Rev.* **D78** 055014 (*Preprint* 0807.0443)
[13] Linares R, Morales-Tecotl H A and Pedraza O 2008 *Phys.Rev.* **D78** 066013 (*Preprint* 0804.2042)

Journal of Physics: Conference Series **545** (2014) 012004 doi:10.1088/1742-6596/545/1/012004

Notoph-Graviton-Photon Coupling

Valeriy V. Dvoeglazov
UAF, Universidad de Zacatecas
Apartado Postal 636, Suc. 3 Cruces, Zacatecas 98068, ZAC., México
URL: http://fisica.uaz.edu.mx/~valeri/

E-mail: valeri@fisica.uaz.edu.mx

Abstract. In the sixties Ogievetskiĭ and Polubarinov proposed the concept of *notoph*, whose helicity properties are complementary to those of *photon*. Later, Kalb and Ramond (and others) developed this theoretical concept. And, at the present times it is widely accepted. We analyze the quantum theory of antisymmetric tensor fields with taking into account mass dimensions of notoph and photon. It appears to be possible the description of both photon and notoph degrees of freedom on the basis of the modified Bargmann-Wigner formalism for the symmetric second-rank spinor.

Next, we proceed to derive equations for the symmetric tensor of the second rank on the basis of the Bargmann-Wigner formalism in a straightforward way. The symmetric multispinor of the fourth rank is used. It is constructed out of the Dirac 4-spinors. Due to serious problems with the interpretation of the results obtained on using the standard procedure we generalize it, and we obtain the spin-2 relativistic equations, which are consistent with the general relativity. The importance of the 4-vector field (and its gauge part) is pointed out.

Thus, we present the full theory which contains the photon, the notoph (the Kalb-Ramond field) and the graviton. The relations of this theory with the higher spin theories are established. In fact, we deduced the gravitational field equations from relativistic quantum mechanics. The relations of this theory with scalar-tensor theories of gravitation and f(R) are discussed. We estimate possible interactions, fermion-notoph, graviton-notoph, photon-notoph, and we conclude that they will be probably seen in experiments in the next few years.

PACS number: 03.65.Pm , 04.50.-h , 11.30.Cp

1. Introduction.

In the series of the papers [1, 2, 3, 4, 5], cf. with Refs. [6, 7, 8], we tried to find connection between the theory of the quantized antisymmetric tensor (AST) field of the second rank (and that of the corresponding 4-vector field) with the $2(2s+1)$ Weinberg-Tucker-Hammer formalism [9, 10].

Several previously published works [11, 12, 13, 14, 15, 16], introduced the concept of the notoph (the Kalb-Ramond field) which is constructed on the basis of the antisymmetric tensor "potentials". It represents itself the non-trivial spin-0 field. The well-known textbooks [17, 18, 19, 20] did *not* discuss the problems, whether the massless quantized AST field and the quantized 4-vector field are transverse or longitudinal fields (in the sense if the helicity $h = \pm 1$ or $h = 0$)? can the electromagnetic potential be a 4-vector in a quantized theory (cf. Ref. [9b,p.251])? how should the massless limit be taken? and many other fundamental problems of the physics of bosons. In my opinion, the most rigorous works are refs. [22, 9, 23, 21], but it is not easy to extract corresponding answers even from them.

First of all, we note that 1) "...In natural units ($c = \hbar = 1$) ... a lagrangian density, since the action is dimensionless, has dimension of $[\text{energy}]^4$"; 2) One can always renormalize the lagrangian density and "one can obtain the same equations of motion... by substituting $L \to (1/M^N)L$, where M is an arbitrary energy scale", cf. [2]; 3) the right physical dimension of the field strength tensor $F^{\mu\nu}$ is $[\text{energy}]^2$; "the transformation $F^{\mu\nu} \to (1/2M)F^{\mu\nu}$ [which was regarded in Ref. [5]] ... requires a more detailed study ... [because] the transformation above changes its physical dimension: it is not a simple normalization transformation". Furthermore, in the first papers on the notoph [12, 13, 14][1] the authors used the normalization of the 4-vector F^μ field[2] to $[\text{energy}]^2$ and, hence, the antisymmetric tensor "potentials" $A^{\mu\nu}$, to $[\text{energy}]^1$. We try to discuss these problems on the basis of the generalized Bargmann-Wigner formalism [22]. Thus, the Proca and Maxwell formalisms are generalized, see, e. g., Ref. [24].

In the Sections 3 and 4 we consider the spin-2 equations. The general scheme for derivation of higher-spin equations has been given in [22]. A field of the rest mass m and the spin $s \geq \frac{1}{2}$ is represented by a completely symmetric multispinor of rank $2s$. The particular cases $s = 1$ and $s = \frac{3}{2}$ have been considered in the textbooks, e. g., Ref. [17]. The spin-2 case can also be of some interest because we can believe that the essential features of the gravitational field are obtained from transverse components of the $(2,0) \oplus (0,2)$ representation of the Lorentz group. Nevertheless, questions of the redundant components of the higher-spin relativistic equations are not yet understood in detail [25].

In the last Sections we discuss the questions of interactions.

2. Photon-Notoph Equations.

For spin 1 we start from[3]

$$[\gamma_{\alpha\beta}p_\alpha p_\beta + Ap_\alpha p_\alpha + Bm^2]\Psi = 0\,, \tag{1}$$

where $p_\mu = -i\partial_\mu$ and $\gamma_{\alpha\beta}$ are the Barut-Muzinich-Williams covariantly - defined 6×6 matrices, $\sum_\mu \gamma_{\mu\mu} = 0$. The determinant of $[\gamma_{\alpha\beta}p_\alpha p_\beta + Ap_\alpha p_\alpha + Bm^2]$ is of the 12th order in p_μ. If we are interested in solutions with $E^2 - \mathbf{p}^2 = m^2$, $c = \hbar = 1$, they can be obtained on using the constraints in the above equation:

$$\frac{B}{A+1} = 1\,, \quad \frac{B}{A-1} = 1\,. \tag{2}$$

We may also have the tachyonic solutions, etc. The particular cases are:

- $A = 0, B = 1 \Leftrightarrow$ we have the Weinberg's equation for $s = 1$ with 3 solutions $E = +\sqrt{\mathbf{p}^2 + m^2}$, 3 solutions $E = -\sqrt{\mathbf{p}^2 + m^2}$, 3 solutions $E = +\sqrt{\mathbf{p}^2 - m^2}$ and 3 solutions $E = -\sqrt{\mathbf{p}^2 - m^2}$. Tachyonic solutions have been reformulated in various ways, for instance, as the ones leading to the spontaneous symmetry breaking, and to the non-zero quantum vacuum.
- $A = 1, B = 2 \Leftrightarrow$ we have the Tucker-Hammer equation for $s = 1$. The solutions are with $E = \pm\sqrt{\mathbf{p}^2 + m^2}$ *only*.

Thus, the addition of the Klein-Gordon equation to (1) may change the physical content even on the *free* level.

[1] It is also known as the longitudinal Kalb-Ramond field, but the consideration of Ogievetskiĭ and Polubarinov permits to study the $m \to 0$ procedure.

[2] It is well known that it is related to the third-rank antisymmetric field tensor.

[3] In the classic works on this formalism the authors worked in the Euclidean metrics. However, there is no any problem to write the equations and other formulas in the pseudo-Euclidean metrics accustomed today; just change the sign of $p_\mu p_\mu$, and other products.

Journal of Physics: Conference Series **545** (2014) 012004 doi:10.1088/1742-6596/545/1/012004

What are the corresponding equations for the antisymmetric tensor field? They can be the Proca equations in the massive case, and the Maxwell equations in the massless case. We have shown in Refs. [1, 2] that one can obtain *four* different equations for antisymmetric tensor fields from the Weinberg's $2(2s+1)-$ component formalism. First of all, we note that Ψ is, in fact, bivector, $\mathbf{E}_i = -iF_{4i}$, $\mathbf{B}_i = \frac{1}{2}\epsilon_{ijk}F_{jk}$,, or $\mathbf{E}_i = -\frac{1}{2}\epsilon_{ijk}\tilde{F}_{jk}$, $\mathbf{B}_i = -i\tilde{F}_{4i}$, or their combinations. One can separate the four cases:

- $\Psi^{(I)} = \begin{pmatrix}\mathbf{E}+i\mathbf{B}\\ \mathbf{E}-i\mathbf{B}\end{pmatrix}$, $P=-1$, where $\mathbf{E}_i$ and $\mathbf{B}_i$ are the components of the tensor.
- $\Psi^{(II)} = \begin{pmatrix}\mathbf{B}-i\mathbf{E}\\ \mathbf{B}+i\mathbf{E}\end{pmatrix}$, $P=+1$, where $\mathbf{E}_i$, $\mathbf{B}_i$ are the components of the tensor.
- $\Psi^{(III)} = \Psi^{(I)}$, but (!) $\mathbf{E}_i$ and $\mathbf{B}_i$ are the corresponding vector and axial-vector components of the *dual* tensor $\tilde{F}_{\mu\nu}$.
- $\Psi^{(IV)} = \Psi^{(II)}$, where $\mathbf{E}_i$ and $\mathbf{B}_i$ are the components of the *dual* tensor $\tilde{F}_{\mu\nu}$.

The mappings of the WTH equations are:

$$\partial_\alpha\partial_\mu F^{(I)}_{\mu\beta} - \partial_\beta\partial_\mu F^{(I)}_{\mu\alpha} + \frac{A-1}{2}\partial_\mu\partial_\mu F^{(I)}_{\alpha\beta} - \frac{B}{2}m^2 F^{(I)}_{\alpha\beta} = 0\,, \tag{3}$$

$$\partial_\alpha\partial_\mu F^{(II)}_{\mu\beta} - \partial_\beta\partial_\mu F^{(II)}_{\mu\alpha} - \frac{A+1}{2}\partial_\mu\partial_\mu F^{(II)}_{\alpha\beta} + \frac{B}{2}m^2 F^{(II)}_{\alpha\beta} = 0\,, \tag{4}$$

$$\partial_\alpha\partial_\mu \tilde{F}^{(III)}_{\mu\beta} - \partial_\beta\partial_\mu \tilde{F}^{(III)}_{\mu\alpha} - \frac{A+1}{2}\partial_\mu\partial_\mu \tilde{F}^{(III)}_{\alpha\beta} + \frac{B}{2}m^2 \tilde{F}^{(III)}_{\alpha\beta} = 0\,, \tag{5}$$

$$\partial_\alpha\partial_\mu \tilde{F}^{(IV)}_{\mu\beta} - \partial_\beta\partial_\mu \tilde{F}^{(IV)}_{\mu\alpha} + \frac{A-1}{2}\partial_\mu\partial_\mu \tilde{F}^{(IV)}_{\alpha\beta} - \frac{B}{2}m^2 \tilde{F}^{(IV)}_{\alpha\beta} = 0\,. \tag{6}$$

In the Tucker-Hammer case ($A=1, B=2$) we can recover the Proca theory from (3):

$$\partial_\alpha\partial_\mu F_{\mu\beta} - \partial_\beta\partial_\mu F_{\mu\alpha} = m^2 F_{\alpha\beta}\,, \tag{7}$$

($A_\nu = \frac{1}{m^2}\partial_\alpha F_{\alpha\nu}$ should be substituted in $F_{\mu\nu} = \partial_\mu A_\nu - \partial_\nu A_\mu$, and the result is multiplied by m^2).

We also noted that the massless limit of this theory does *not* coincide with the Maxwell theory in some cases, while it contains the latter as a particular case. In [3, 5, 30] we showed that it is possible to define various massless limits for the Duffin-Kemmer-Proca theory. Another one is the Ogievetskiĭ-Polubarinov *notoph* (which is also called the Kalb-Ramond field), Ref. [12] in the US literature. The transverse components of the AST field can be removed from the corresponding Lagrangian by means of the "new gauge transformation" $A_{\mu\nu} \to A_{\mu\nu} + \partial_\mu\Lambda_\nu - \partial_\nu\Lambda_\mu$, with the vector gauge function Λ_μ.

The second (II) case is

$$\partial_\alpha\partial_\mu F_{\mu\beta} - \partial_\beta\partial_\mu F_{\mu\alpha} = [\partial_\mu\partial_\mu - m^2]F_{\alpha\beta}\,. \tag{8}$$

So, on the mass shell we have $[\partial_\mu\partial_\mu - m^2]F_{\alpha\beta} = 0$, and, hence,

$$\partial_\alpha\partial_\mu F_{\mu\beta} - \partial_\beta\partial_\mu F_{\mu\alpha} = 0\,, \tag{9}$$

which rather corresponds to the Maxwell-like case. However, if we calculate dispersion relations for the second case, Eq. (9), it appears that the equation has solutions even if $m \neq 0$.

Now we are interested in the *parity-violating* equations for antisymmetric tensor fields. We investigate the most general mapping of the Weinberg-Tucker-Hammer formulation to the antisymetric tensor field formulation too. Instead of $\Psi^{(I-IV)}$ we shall try to use now

$$\Psi^{(A)} = \begin{pmatrix} \mathbf{E}+i\mathbf{B} \\ \mathbf{B}+i\mathbf{E} \end{pmatrix} = \frac{1+\gamma^5}{2}\Psi^{(I)} + \frac{1-\gamma^5}{2}\Psi^{(II)} . \tag{10}$$

As a result, the equation for the AST fields is

$$\partial_\alpha\partial_\mu F_{\mu\beta} - \partial_\beta\partial_\mu F_{\mu\alpha} = \frac{1}{2}(\partial_\mu\partial_\mu)F_{\alpha\beta} + [-\frac{A}{2}(\partial_\mu\partial_\mu) + \frac{B}{2}m^2]\tilde{F}_{\alpha\beta} . \tag{11}$$

Of course, $\Psi^{(A)\prime} = \begin{pmatrix} \mathbf{B}-i\mathbf{E} \\ \mathbf{E}-i\mathbf{B} \end{pmatrix} = -i\Psi^{(A)}$, and the equation is unchanged. The different choice is

$$\Psi^{(B)} = \begin{pmatrix} \mathbf{E}+i\mathbf{B} \\ -\mathbf{B}-i\mathbf{E} \end{pmatrix} = \frac{1+\gamma^5}{2}\Psi^{(I)} - \frac{1-\gamma^5}{2}\Psi^{(II)} . \tag{12}$$

Thus, one has

$$\partial_\alpha\partial_\mu F_{\mu\beta} - \partial_\beta\partial_\mu F_{\mu\alpha} = \frac{1}{2}(\partial_\mu\partial_\mu)F_{\alpha\beta} + [\frac{A}{2}(\partial_\mu\partial_\mu) - \frac{B}{2}m^2]\tilde{F}_{\alpha\beta} . \tag{13}$$

Of course, one can also use the dual tensor ($\mathbf{E}^i = -\frac{1}{2}\epsilon_{ijk}\tilde{F}_{jk}$ and $\mathbf{B}^i = -i\tilde{F}_{4i}$) and obtain analogous equations:

$$\partial_\alpha\partial_\mu \tilde{F}_{\mu\beta} - \partial_\beta\partial_\mu \tilde{F}_{\mu\alpha} = \frac{1}{2}(\partial_\mu\partial_\mu)\tilde{F}_{\alpha\beta} + [-\frac{A}{2}(\partial_\mu\partial_\mu) + \frac{B}{2}m^2]F_{\alpha\beta} , \tag{14}$$

$$\partial_\alpha\partial_\mu \tilde{F}_{\mu\beta} - \partial_\beta\partial_\mu \tilde{F}_{\mu\alpha} = \frac{1}{2}(\partial_\mu\partial_\mu)\tilde{F}_{\alpha\beta} + [\frac{A}{2}(\partial_\mu\partial_\mu) - \frac{B}{2}m^2]F_{\alpha\beta} . \tag{15}$$

They are connected with (11,13) by the dual transformations.

The states corresponding to the new functions $\Psi^{(A)}$, $\Psi^{(B)}$ etc are *not* the parity eigenstates. So, it is not surprising that we have $F_{\alpha\beta}$ and its dual $\tilde{F}_{\alpha\beta}$ in the same equations. In total we have already eight equations.

One can also consider the most general case

$$\Psi^{(W)} = \begin{pmatrix} aF_{4i} + b\tilde{F}_{4i} + c\epsilon_{ijk}F_{jk} + d\epsilon_{ijk}\tilde{F}_{jk} \\ eF_{4i} + f\tilde{F}_{4i} + g\epsilon_{ijk}F_{jk} + h\epsilon_{ijk}\tilde{F}_{jk} \end{pmatrix} . \tag{16}$$

So, we shall have dynamical equations for $F_{\alpha\beta}$ and $\tilde{F}_{\alpha\beta}$ with additional parameters $a, b, c, d, \ldots \in \mathbf{C}$. We have a lot of antisymmetric tensor fields here. However,

- the covariant form preserves if there are some restrictions on the parameters, only. Alternatively, we have some additional terms of ∂_4^2 or ∇^2;
- both $F_{\mu\nu}$ and $\tilde{F}_{\mu\nu}$ are present in the equations;
- under the definite choice of $a, b, c, d \ldots$ the equations can be reduced to the above equations for the tensor $\mathcal{H}_{\mu\nu}$ and its dual:

$$\mathcal{H}_{\mu\nu} = c_1 F_{\mu\nu} + c_2\tilde{F}_{\mu\nu} + \frac{c_3}{2}\epsilon_{\mu\nu\alpha\beta}F_{\alpha\beta} + \frac{c_4}{2}\epsilon_{\mu\nu\alpha\beta}\tilde{F}_{\alpha\beta} ; \tag{17}$$

Journal of Physics: Conference Series **545** (2014) 012004 doi:10.1088/1742-6596/545/1/012004

- the parity properties of $\Psi^{(W)}$ are very complicated.

Anther way of constructing the equations of high-spin particles has been given in [22, 17].[4] Bargmann and Wigner claimed explicitly that they constructed $(2s+1)$ states.[5] Below we present the standard Bargmann-Wigner formalism for the spin $s=1$ (and turn to the standard pseudo-Euclidean metric):

$$[i\gamma^\mu \partial_\mu - m]_{\alpha\beta} \Psi_{\beta\gamma} = 0\,, \tag{18}$$
$$[i\gamma^\mu \partial_\mu - m]_{\gamma\beta} \Psi_{\alpha\beta} = 0\,, \tag{19}$$

If one has

$$\Psi_{\{\alpha\beta\}} = (\gamma^\mu R)_{\alpha\beta} A_\mu + (\sigma^{\mu\nu} R)_{\alpha\beta} F_{\mu\nu}\,, \tag{20}$$

with[6]

$$R = e^{i\varphi} \begin{pmatrix} \Theta & 0 \\ 0 & -\Theta \end{pmatrix} \quad \Theta = \begin{pmatrix} 0 & -1 \\ 1 & 0 \end{pmatrix} \tag{25}$$

in the spinorial representation of γ-matrices, we obtain the Duffin-Kemmer-Proca equations:

$$\partial^\alpha F_{\alpha\mu} = \frac{m}{2} A_\mu\,, \tag{26}$$
$$2m F_{\mu\nu} = \partial_\mu A_\nu - \partial_\nu A_\mu\,. \tag{27}$$

In order to obtain these equations one should add the equations (18,19) and compare functional coefficients at the corresponding commutators, see Ref. [17]. After the corresponding renormalization $A_\mu \to 2mA_\mu$ (or $F^{\mu\nu} \to (1/2m) F^{\mu\nu}$), we obtain the standard textbook set:

$$\partial^\alpha F_{\alpha\mu} = m^2 A_\mu\,, \tag{28}$$
$$F_{\mu\nu} = \partial_\mu A_\nu - \partial_\nu A_\mu\,. \tag{29}$$

It gives the equation (7) for the antisymmetric tensor field. Of course, one can investigate other sets of equations with different normalization of the $F_{\mu\nu}$ and A_μ fields. Are all these sets of equations equivalent? As we see, to answer this question is not trivial. It was argued that the physical normalization is such that in the massless limit the zero-momentum field functions should vanish in the momentum representation (there are no massless particles at rest). Moreover, we advocate the following approach: the massless limit can and must be taken in the end of all calculations only, i. e., for physical quantities.

How can one obtain other equations following from the Weinberg-Tucker-Hammer approach? The recipe for the third equation is simple: use, instead of $(\sigma^{\mu\nu}R)F_{\mu\nu}$, another symmetric matrix $(\gamma^5 \sigma^{\mu\nu} R) F_{\mu\nu}$.

[4] On can also obtain the $s=0$ Kemmer equations on using the Bargmann-Wigner procedure. One should use the *antisymmetric* second-rank multispinor in this case.

[5] The Weinberg-Tucker-Hammer theory has essentially $2(2s+1)$ components.

[6] The reflection operator R has the properties

$$R^T = -R\,, \quad R^\dagger = R = R^{-1}\,, \tag{21}$$
$$R^{-1}\gamma^5 R = (\gamma^5)^T\,, \tag{22}$$
$$R^{-1}\gamma^\mu R = -(\gamma^\mu)^T\,, \tag{23}$$
$$R^{-1}\sigma^{\mu\nu} R = -(\sigma^{\mu\nu})^T\,. \tag{24}$$

After taking into account the above observations let us repeat the procedure of derivation of the Proca equations from the Bargmann-Wigner equations for a *symmetric* second-rank spinor. However, we now use

$$\Psi_{\{\alpha\beta\}} = (\gamma^\mu R)_{\alpha\beta}(c_a m A_\mu + c_f F_\mu) + (\sigma^{\mu\nu} R)_{\alpha\rho}(c_A m(\gamma^5)_{\rho\beta} A_{\mu\nu} + c_F I_{\rho\beta} F_{\mu\nu})\,, \tag{30}$$

with the same R and Θ as above. Matrices γ^μ are again chosen in the Weyl (spinorial) representation, i.e., γ^5 is assumed to be diagonal. Constants c_i are some numerical dimensionless coefficients. The properties of the reflection operator R are necessary for the expansion (30) to be possible in such a form, i.e., in order to have the $\gamma^\mu R$, $\sigma^{\mu\nu} R$ and $\gamma^5\sigma^{\mu\nu} R$ to be *symmetric* matrices.

The substitution of the above expansion into the Bargmann-Wigner equations, Ref. [17], gives us the new Proca-like equations:

$$c_a m(\partial_\mu A_\nu - \partial_\nu A_\mu) + c_f(\partial_\mu F_\nu - \partial_\nu F_\mu) = i c_A m^2 \epsilon_{\alpha\beta\mu\nu} A^{\alpha\beta} + 2 m c_F F_{\mu\nu}\,, \tag{31}$$

$$c_a m^2 A_\mu + c_f m F_\mu = i c_A m \epsilon_{\mu\nu\alpha\beta} \partial^\nu A^{\alpha\beta} + 2 c_F \partial^\nu F_{\mu\nu}\,. \tag{32}$$

In the case $c_a = 1$, $c_F = \frac{1}{2}$ and $c_f = c_A = 0$ they are reduced to the ordinary Proca equations.[7] In the general case we obtain dynamical equations which connect the photon, the notoph and their potentials. The divergent (in $m \to 0$) parts of field functions and those of dynamical variables should be removed by corresponding gauge (or Kalb-Ramond gauge) transformations. It is well known that the notoph massless field is considered to be the pure longitudinal field after one takes into account $\partial_\mu A^{\mu\nu} = 0$. Apart from these dynamical equations we can obtain a number of constraints by means of the subtraction of the equations of the Bargmann-Wigner system (instead of the addition as for (31,32)). They read

$$m c_a \partial^\mu A_\mu + c_f \partial^\mu f_\mu = 0\,, \tag{33}$$

$$m c_A \partial^\alpha A_{\alpha\mu} + \frac{i}{2} c_F \epsilon_{\alpha\beta\nu\mu} \partial^\alpha F^{\beta\nu} = 0, \tag{34}$$

that suggests $\widetilde{F}^{\mu\nu} \sim i m A^{\mu\nu}$ and $f^\mu \sim m A^\mu$, as in [12].

Thus, after the suitable choice of the dimensionless coefficients c_i the Lagrangian density for the photon-notoph field can be proposed:

$$\begin{aligned}\mathcal{L} &= \mathcal{L}^{Proca} + \mathcal{L}^{Notoph} = -\frac{1}{8} F_\mu F^\mu - \frac{1}{4} F_{\mu\nu} F^{\mu\nu} + \\ &+ \frac{m^2}{2} A_\mu A^\mu + \frac{m^2}{4} A_{\mu\nu} A^{\mu\nu}\,,\end{aligned} \tag{35}$$

The limit $m \to 0$ may be taken for dynamical variables, in the end of calculations only.

Furthermore, it is logical to introduce the normalization scalar field $\varphi(x)$, and consider the expansion:

$$\Psi_{\{\alpha\beta\}} = (\gamma^\mu R)_{\alpha\beta}(\varphi A_\mu) + (\sigma^{\mu\nu} R)_{\alpha\beta} F_{\mu\nu}\,. \tag{36}$$

Then, we arrive at the following set

$$2 m F_{\mu\nu} = \varphi(\partial_\mu A_\nu - \partial_\nu A_\mu) + (\partial_\mu \varphi) A_\nu - (\partial_\nu \varphi) A_\mu\,, \tag{37}$$

$$\partial^\nu F_{\mu\nu} = \frac{m}{2}(\varphi A_\mu)\,, \tag{38}$$

[7] We still note that the division by m in the first equation is *not* the well-defined operation in the case if someone is interested in the subsequent limiting procedure $m \to 0$. Probably, in order to avoid this obscure point one may wish to write the Dirac equations in the form $[(i\gamma^\mu \partial_\mu)/m - I]\,\psi(x) = 0$, which follows straightforwardly in the derivation of the Dirac equation on the basis of the Ryder relation [7] and the Wigner rules for the boosts of the field functions from the zero-momentum frame.

Journal of Physics: Conference Series **545** (2014) 012004 doi:10.1088/1742-6596/545/1/012004

which in the case of the constant scalar field $\varphi = 2m$ can also be reduced to the system of the Proca equations. The additional constraints are

$$(\partial^\mu \varphi) A_\mu + \varphi(\partial^\mu A_\mu) = 0\,, \tag{39}$$

$$\partial_\mu \widetilde{F}^{\mu\nu} = 0\,. \tag{40}$$

At the moment it is not yet obvious, how can we account for other equations in the $(1,0)\oplus(0,1)$ representation, e.g. [7b], rigorously. For instance, one can wish to seek the generalization of the Proca equations on the basis of the introduction of two mass parameters m_1 and m_2. But, when we apply the BW procedure to the Dirac equations we cannot obtain new physical content. Another equation in the $(1/2,0)\oplus(0,1/2)$ representation was discussed in Ref. [26]. It has the form:

$$\left[i\gamma^\mu \partial_\mu - m_1 - \gamma^5 m_2\right] \Psi(x) = 0\,. \tag{41}$$

The Bargmann-Wigner procedure for this system of equations (which include the γ^5 matrix in the mass term) yields:

$$2m_1 F^{\mu\nu} + 2im_2 \widetilde{F}^{\mu\nu} = \varphi(\partial^\mu A^\nu - \partial^\nu A^\mu) + (\partial^\mu\varphi)A^\nu - (\partial^\nu\varphi)A^\mu\,, \tag{42}$$

$$\partial^\nu F_{\mu\nu} = \frac{m_1}{2}(\varphi A_\mu), \tag{43}$$

with the constraints

$$(\partial^\mu \varphi) A_\mu + \varphi(\partial^\mu A_\mu) = 0\,, \tag{44}$$

$$\partial^\nu \widetilde{F}_{\mu\nu} = \frac{im_2}{2}(\varphi A_\mu)\,. \tag{45}$$

In general, we can now use the four different mass parameters in the equations which are analogous to (18,19). However, the equality of mass factors[8] ($m_1^{(1)} = m_1^{(2)}$ and $m_2^{(1)} = m_2^{(2)}$) is obtained as necessary conditions in the process of calculations in the system of the Dirac-like equations.

In fact, the results of this paper develop the old results of Ref. [12]. According to [12, Eqs.(9,10)] we proceed in constructing the "potentials" for the notoph as follows:[9]

$$A_{\mu\nu}(\mathbf{p}) = N\left[\epsilon_\mu^{(1)}(\mathbf{p})\epsilon_\nu^{(2)}(\mathbf{p}) - \epsilon_\nu^{(1)}(\mathbf{p})\epsilon_\mu^{(2)}(\mathbf{p})\right]\,. \tag{46}$$

We use explicit forms for the polarization vectors (e.g., Refs. [21] and [5, formulas(15a,b)]) boosted to the momentum $\mathbf{p}$:

$$\epsilon^\mu(\mathbf{0},+1) = -\frac{1}{\sqrt{2}}\begin{pmatrix}0\\1\\i\\0\end{pmatrix}, \epsilon^\mu(\mathbf{0},0) = \begin{pmatrix}0\\0\\0\\1\end{pmatrix}, \epsilon^\mu(\mathbf{0},-1) = \frac{1}{\sqrt{2}}\begin{pmatrix}0\\1\\-i\\0\end{pmatrix}, \tag{47}$$

and ($\widehat{p}_i = p_i/|\mathbf{p}|$, $\gamma = E_p/m$), Ref. [21, p.68] or Ref. [19, p.108],

$$\epsilon^\mu(\mathbf{p},\sigma) = L^\mu{}_\nu(\mathbf{p})\epsilon^\nu(\mathbf{0},\sigma)\,, \tag{48}$$

$$L^0{}_0(\mathbf{p}) = \gamma\,,\quad L^i{}_0(\mathbf{p}) = L^0{}_i(\mathbf{p}) = \widehat{p}_i\sqrt{\gamma^2-1}\,, \tag{49}$$

$$L^i{}_k(\mathbf{p}) = \delta_{ik} + (\gamma-1)\widehat{p}_i\widehat{p}_k\,. \tag{50}$$

[8] Here, the superscripts (1) and (2) refers to the first and the second equations, respectively, in the modified Bargmann-Wigner system.

[9] The notation is that of Ref. [12] here.

N, the normalization factor, should be taken into account for possible analyses of propagators and massless limits. After substitutions in the definition (46) one obtains

$$A^{\mu\nu}(\mathbf{p}) = \frac{iN^2}{m}\begin{pmatrix} 0 & -p_2 & p_1 & 0 \\ p_2 & 0 & m+\frac{p_r p_l}{p_0+m} & \frac{p_2 p_3}{p_0+m} \\ -p_1 & -m-\frac{p_r p_l}{p_0+m} & 0 & -\frac{p_1 p_3}{p_0+m} \\ 0 & -\frac{p_2 p_3}{p_0+m} & \frac{p_1 p_3}{p_0+m} & 0 \end{pmatrix}, \tag{51}$$

i.e., it coincides with the longitudinal components of the antisymmetric tensor obtained in Refs. [7a,Eqs.(2.14,2.17)] and [5, Eqs.(17b,18b)] within the normalization and different forms of the spin basis. The $A_{\mu\nu}(\mathbf{p})$ potential reduces to zero in the limiting case ($m \to 0$) under appropriate choice of the normalization $N = m^\alpha$, $\alpha > 1/2$. If $N = \sqrt{m}$ this reduction of the non-transverse state occurs if a $s = 1$ particle moves along with the third axis OZ.[10] It is also useful to compare Eq. (51) with the formula (B2) in Ref. [8] in order to think about correct procedures for taking the massless limits.

Next, the Tam-Happer experiments [27] on two laser beams interaction did not find satisfactory explanation in the framework of the ordinary QED (at least, their explanation is complicated by huge technical calculations). On the other hand, in Ref. [28] a very interesting model has been proposed. It is based on gauging the Dirac field on using the coordinate-dependent parameters $\alpha_{\mu\nu}(x)$ in

$$\psi(x) \to \psi'(x') = \Omega\psi(x)\,, \quad \Omega = \exp\left[\frac{i}{2}\sigma^{\mu\nu}\alpha_{\mu\nu}(x)\right]\,, \tag{52}$$

and, thus, the second "photon" was introduced. The compensating 24-component (in general) field $B_{\mu,\nu\lambda}$ reduces to the 4-vector field as follows (the notation of [28] is used here):

$$B_{\mu,\nu\lambda} = \frac{1}{4}\epsilon_{\mu\nu\lambda\sigma}a_\sigma(x)\,. \tag{53}$$

As readily seen, after comparison of these formulas with those of Refs. [12, 13, 14], the second photon is nothing more than the Ogievetskiĭ-Polubarinov *notoph* within the normalization. Parity properties are dependent not only on the explicit forms of the momentum-space field functions of the (1/2, 1/2) representation, but also on the properties of corresponding creation/annihilation operators. Helicity properties depend on the normalization.

3. The Standard Bargmann-Wigner Formalism Applied for Spin 2.

In this Section we use the commonly-accepted procedure for the derivation of higher-spin equations [22]. We begin with the equations for the 4-rank symmetric spinor:

$$[i\gamma^\mu\partial_\mu - m]_{\alpha\alpha'}\Psi_{\alpha'\beta\gamma\delta} = 0\,, \tag{54}$$
$$[i\gamma^\mu\partial_\mu - m]_{\beta\beta'}\Psi_{\alpha\beta'\gamma\delta} = 0\,, \tag{55}$$
$$[i\gamma^\mu\partial_\mu - m]_{\gamma\gamma'}\Psi_{\alpha\beta\gamma'\delta} = 0\,, \tag{56}$$
$$[i\gamma^\mu\partial_\mu - m]_{\delta\delta'}\Psi_{\alpha\beta\gamma\delta'} = 0\,. \tag{57}$$

The massless limit (if one needs) should be taken in the end of all calculations.

We proceed expanding the field function in the set of symmetric matrices (as in the spin-1 case, cf. Ref. [5]). In the beginning let us use the first two indices:[11]

$$\Psi_{\{\alpha\beta\}\gamma\delta} = (\gamma_\mu R)_{\alpha\beta}\Psi^\mu_{\gamma\delta} + (\sigma_{\mu\nu}R)_{\alpha\beta}\Psi^{\mu\nu}_{\gamma\delta}\,. \tag{58}$$

[10] But, even in this case we cannot have a good behaviour of the 4-vector fields/potentials in the massless limit in the instant form of the relativistic dynamics, cf. [8].

[11] The matrix R can be related to the CP operation in the $(1/2, 0) \oplus (0, 1/2)$ representation.

We would like to write the corresponding equations for functions $\Psi^{\mu}_{\gamma\delta}$ and $\Psi^{\mu\nu}_{\gamma\delta}$ in the form:

$$\frac{2}{m}\partial_\mu \Psi^{\mu\nu}_{\gamma\delta} = -\Psi^{\nu}_{\gamma\delta}\,, \tag{59}$$

$$\Psi^{\mu\nu}_{\gamma\delta} = \frac{1}{2m}\left[\partial^\mu \Psi^{\nu}_{\gamma\delta} - \partial^\nu \Psi^{\mu}_{\gamma\delta}\right]. \tag{60}$$

Constraints $(1/m)\partial_\mu \Psi^{\mu}_{\gamma\delta} = 0$ and $(1/m)\epsilon^{\mu\nu}{}_{\alpha\beta}\,\partial_\mu \Psi^{\alpha\beta}_{\gamma\delta} = 0$ can be regarded as the consequence of Eqs. (59,60).

Next, we present the vector-spinor and tensor-spinor functions as

$$\Psi^{\mu}_{\{\gamma\delta\}} = (\gamma^\kappa R)_{\gamma\delta} G_\kappa{}^{\mu} + (\sigma^{\kappa\tau} R)_{\gamma\delta} F_{\kappa\tau}{}^{\mu}\,, \tag{61}$$

$$\Psi^{\mu\nu}_{\{\gamma\delta\}} = (\gamma^\kappa R)_{\gamma\delta} T_\kappa{}^{\mu\nu} + (\sigma^{\kappa\tau} R)_{\gamma\delta} R_{\kappa\tau}{}^{\mu\nu}\,, \tag{62}$$

i. e., using the symmetric matrix coefficients in indices γ and δ. Hence, the total function is

$$\begin{aligned}\Psi_{\{\alpha\beta\}\{\gamma\delta\}} &= (\gamma_\mu R)_{\alpha\beta}(\gamma^\kappa R)_{\gamma\delta} G_\kappa{}^{\mu} + (\gamma_\mu R)_{\alpha\beta}(\sigma^{\kappa\tau} R)_{\gamma\delta} F_{\kappa\tau}{}^{\mu} + \\ &+ (\sigma_{\mu\nu} R)_{\alpha\beta}(\gamma^\kappa R)_{\gamma\delta} T_\kappa{}^{\mu\nu} + (\sigma_{\mu\nu} R)_{\alpha\beta}(\sigma^{\kappa\tau} R)_{\gamma\delta} R_{\kappa\tau}{}^{\mu\nu}\,,\end{aligned} \tag{63}$$

and the resulting tensor equations are:

$$\frac{2}{m}\partial_\mu T_\kappa{}^{\mu\nu} = -G_\kappa{}^{\nu}\,, \tag{64}$$

$$\frac{2}{m}\partial_\mu R_{\kappa\tau}{}^{\mu\nu} = -F_{\kappa\tau}{}^{\nu}\,, \tag{65}$$

$$T_\kappa{}^{\mu\nu} = \frac{1}{2m}\left[\partial^\mu G_\kappa{}^{\nu} - \partial^\nu G_\kappa{}^{\mu}\right], \tag{66}$$

$$R_{\kappa\tau}{}^{\mu\nu} = \frac{1}{2m}\left[\partial^\mu F_{\kappa\tau}{}^{\nu} - \partial^\nu F_{\kappa\tau}{}^{\mu}\right]. \tag{67}$$

The constraints are re-written to

$$\frac{1}{m}\partial_\mu G_\kappa{}^{\mu} = 0\,, \quad \frac{1}{m}\partial_\mu F_{\kappa\tau}{}^{\mu} = 0\,, \tag{68}$$

$$\frac{1}{m}\epsilon_{\alpha\beta\nu\mu}\partial^\alpha T_\kappa{}^{\beta\nu} = 0\,, \quad \frac{1}{m}\epsilon_{\alpha\beta\nu\mu}\partial^\alpha R_{\kappa\tau}{}^{\beta\nu} = 0\,. \tag{69}$$

However, we need to make symmetrization over these two sets of indices $\{\alpha\beta\}$ and $\{\gamma\delta\}$. The total symmetry can be ensured if one contracts the function $\Psi_{\{\alpha\beta\}\{\gamma\delta\}}$ with *antisymmetric* matrices $R^{-1}_{\beta\gamma}$, $(R^{-1}\gamma^5)_{\beta\gamma}$ and $(R^{-1}\gamma^5\gamma^\lambda)_{\beta\gamma}$, and equate all these contractions to zero (similar to the $s = 3/2$ case considered in Ref. [17, p. 44]. We obtain additional constraints on the tensor field functions:

$$G_\mu{}^{\mu} = 0\,, \quad G_{[\kappa\,\mu]} = 0\,, \quad G^{\kappa\mu} = \frac{1}{2} g^{\kappa\mu} G_\nu{}^{\nu}\,, \tag{70}$$

$$F_{\kappa\mu}{}^{\mu} = F_{\mu\kappa}{}^{\mu} = 0\,, \quad \epsilon^{\kappa\tau\mu\nu} F_{\kappa\tau,\mu} = 0\,, \tag{71}$$

$$T^{\mu}{}_{\mu\kappa} = T^{\mu}{}_{\kappa\mu} = 0\,, \quad \epsilon^{\kappa\tau\mu\nu} T_{\kappa,\tau\mu} = 0\,, \tag{72}$$

$$F^{\kappa\tau,\mu} = T^{\mu,\kappa\tau}\,, \quad \epsilon^{\kappa\tau\mu\lambda}(F_{\kappa\tau,\mu} + T_{\kappa,\tau\mu}) = 0\,, \tag{73}$$

$$R_{\kappa\nu}{}^{\mu\nu} = R_{\nu\kappa}{}^{\mu\nu} = R_{\kappa\nu}{}^{\nu\mu} = R_{\nu\kappa}{}^{\nu\mu} = R_{\mu\nu}{}^{\mu\nu} = 0\,, \tag{74}$$

$$\epsilon^{\mu\nu\alpha\beta}(g_{\beta\kappa} R_{\mu\tau,\nu\alpha} - g_{\beta\tau} R_{\nu\alpha,\mu\kappa}) = 0 \quad \epsilon^{\kappa\tau\mu\nu} R_{\kappa\tau,\mu\nu} = 0\,. \tag{75}$$

Thus, we encountered with the well-known difficulty of the theory of spin-2 particles in the Minkowski space. We explicitly showed that all field functions become to be equal to zero. Such a situation cannot be considered as a satisfactory one (because it does not give us any physical information), and it can be corrected in several ways.[12]

4. The Generalized Bargmann-Wigner Formalism for Spin 2.

We shall modify the formalism in the spirit of Ref. [30]. The field function (58) is now presented as

$$\Psi_{\{\alpha\beta\}\gamma\delta} = \alpha_1(\gamma_\mu R)_{\alpha\beta}\Psi^\mu_{\gamma\delta} + \alpha_2(\sigma_{\mu\nu}R)_{\alpha\beta}\Psi^{\mu\nu}_{\gamma\delta} + \alpha_3(\gamma^5\sigma_{\mu\nu}R)_{\alpha\beta}\widetilde{\Psi}^{\mu\nu}_{\gamma\delta}\,, \tag{76}$$

with

$$\Psi^\mu_{\{\gamma\delta\}} = \beta_1(\gamma^\kappa R)_{\gamma\delta}G_\kappa{}^{\ \mu} + \beta_2(\sigma^{\kappa\tau}R)_{\gamma\delta}F_{\kappa\tau}{}^{\mu} + \beta_3(\gamma^5\sigma^{\kappa\tau}R)_{\gamma\delta}\widetilde{F}_{\kappa\tau}{}^{\mu}\,, \tag{77}$$

$$\Psi^{\mu\nu}_{\{\gamma\delta\}} = \beta_4(\gamma^\kappa R)_{\gamma\delta}T_\kappa{}^{\ \mu\nu} + \beta_5(\sigma^{\kappa\tau}R)_{\gamma\delta}R_{\kappa\tau}{}^{\mu\nu} + \beta_6(\gamma^5\sigma^{\kappa\tau}R)_{\gamma\delta}\widetilde{R}_{\kappa\tau}{}^{\mu\nu}\,, \tag{78}$$

$$\widetilde{\Psi}^{\mu\nu}_{\{\gamma\delta\}} = \beta_7(\gamma^\kappa R)_{\gamma\delta}\widetilde{T}_\kappa{}^{\ \mu\nu} + \beta_8(\sigma^{\kappa\tau}R)_{\gamma\delta}\widetilde{D}_{\kappa\tau}{}^{\mu\nu} + \beta_9(\gamma^5\sigma^{\kappa\tau}R)_{\gamma\delta}D_{\kappa\tau}{}^{\mu\nu}\,. \tag{79}$$

Hence, the function $\Psi_{\{\alpha\beta\}\{\gamma\delta\}}$ can be expressed as a sum of nine terms:

$$\begin{aligned}\Psi_{\{\alpha\beta\}\{\gamma\delta\}} &= \alpha_1\beta_1(\gamma_\mu R)_{\alpha\beta}(\gamma^\kappa R)_{\gamma\delta}G_\kappa{}^{\ \mu} + \alpha_1\beta_2(\gamma_\mu R)_{\alpha\beta}(\sigma^{\kappa\tau}R)_{\gamma\delta}F_{\kappa\tau}{}^{\mu} + \\ &+ \alpha_1\beta_3(\gamma_\mu R)_{\alpha\beta}(\gamma^5\sigma^{\kappa\tau}R)_{\gamma\delta}\widetilde{F}_{\kappa\tau}{}^{\mu} + +\alpha_2\beta_4(\sigma_{\mu\nu}R)_{\alpha\beta}(\gamma^\kappa R)_{\gamma\delta}T_\kappa{}^{\ \mu\nu} + \\ &+ \alpha_2\beta_5(\sigma_{\mu\nu}R)_{\alpha\beta}(\sigma^{\kappa\tau}R)_{\gamma\delta}R_{\kappa\tau}{}^{\mu\nu} + \alpha_2\beta_6(\sigma_{\mu\nu}R)_{\alpha\beta}(\gamma^5\sigma^{\kappa\tau}R)_{\gamma\delta}\widetilde{R}_{\kappa\tau}{}^{\mu\nu} + \\ &+ \alpha_3\beta_7(\gamma^5\sigma_{\mu\nu}R)_{\alpha\beta}(\gamma^\kappa R)_{\gamma\delta}\widetilde{T}_\kappa{}^{\ \mu\nu} + \alpha_3\beta_8(\gamma^5\sigma_{\mu\nu}R)_{\alpha\beta}(\sigma^{\kappa\tau}R)_{\gamma\delta}\widetilde{D}_{\kappa\tau}{}^{\mu\nu} + \\ &+ \alpha_3\beta_9(\gamma^5\sigma_{\mu\nu}R)_{\alpha\beta}(\gamma^5\sigma^{\kappa\tau}R)_{\gamma\delta}D_{\kappa\tau}{}^{\mu\nu}\,.\end{aligned} \tag{80}$$

The corresponding dynamical equations are given by[13]

$$\frac{2\alpha_2\beta_4}{m}\partial_\nu T_\kappa{}^{\ \mu\nu} + \frac{i\alpha_3\beta_7}{m}\epsilon^{\mu\nu\alpha\beta}\partial_\nu\widetilde{T}_{\kappa,\alpha\beta} = \alpha_1\beta_1 G_\kappa{}^{\ \mu}\,, \tag{81}$$

$$\begin{aligned}&\frac{2\alpha_2\beta_5}{m}\partial_\nu R_{\kappa\tau}{}^{\mu\nu} + \frac{i\alpha_2\beta_6}{m}\epsilon_{\alpha\beta\kappa\tau}\partial_\nu\widetilde{R}^{\alpha\beta,\mu\nu} + \frac{i\alpha_3\beta_8}{m}\epsilon^{\mu\nu\alpha\beta}\partial_\nu\widetilde{D}_{\kappa\tau,\alpha\beta} - \\ &- \frac{\alpha_3\beta_9}{2}\epsilon^{\mu\nu\alpha\beta}\epsilon_{\lambda\delta\kappa\tau}D^{\lambda\delta}{}_{\alpha\beta} = \alpha_1\beta_2 F_{\kappa\tau}{}^{\mu} + \frac{i\alpha_1\beta_3}{2}\epsilon_{\alpha\beta\kappa\tau}\widetilde{F}^{\alpha\beta,\mu}\,,\end{aligned} \tag{82}$$

$$2\alpha_2\beta_4 T_\kappa{}^{\ \mu\nu} + i\alpha_3\beta_7\epsilon^{\alpha\beta\mu\nu}\widetilde{T}_{\kappa,\alpha\beta} = \frac{\alpha_1\beta_1}{m}(\partial^\mu G_\kappa{}^{\ \nu} - \partial^\nu G_\kappa{}^{\ \mu})\,, \tag{83}$$

$$\begin{aligned}&2\alpha_2\beta_5 R_{\kappa\tau}{}^{\mu\nu} + i\alpha_3\beta_8\epsilon^{\alpha\beta\mu\nu}\widetilde{D}_{\kappa\tau,\alpha\beta} + i\alpha_2\beta_6\epsilon_{\alpha\beta\kappa\tau}\widetilde{R}^{\alpha\beta,\mu\nu} - \frac{\alpha_3\beta_9}{2}\epsilon^{\alpha\beta\mu\nu}\epsilon_{\lambda\delta\kappa\tau}D^{\lambda\delta}{}_{\alpha\beta} = \\ &= \frac{\alpha_1\beta_2}{m}(\partial^\mu F_{\kappa\tau}{}^{\nu} - \partial^\nu F_{\kappa\tau}{}^{\mu}) + \frac{i\alpha_1\beta_3}{2m}\epsilon_{\alpha\beta\kappa\tau}(\partial^\mu\widetilde{F}^{\alpha\beta,\nu} - \partial^\nu\widetilde{F}^{\alpha\beta,\mu})\,.\end{aligned} \tag{84}$$

The essential constraints are:

$$\alpha_1\beta_1 G^\mu{}_\mu = 0\,, \quad \alpha_1\beta_1 G_{[\kappa\mu]} = 0\,, \tag{85}$$

[12] The reader can compare our results of this Section with those of Ref. [29]. I became aware about their consideration from Dr. D. V. Ahluwalia (personal communications, May 5, 1998). I consider their discussion of the standard formalism in the Sections I and II, as insufficient.

[13] All indices in this formula are already pure vectorial and have nothing to do with previous notation. The coefficients α_i and β_i may, in general, carry some dimension.

$$2i\alpha_1\beta_2 F_{\alpha\mu}{}^{\mu} + \alpha_1\beta_3\epsilon^{\kappa\tau\mu}{}_{\alpha}\widetilde{F}_{\kappa\tau,\mu} = 0\,, \tag{86}$$

$$2i\alpha_1\beta_3 \widetilde{F}_{\alpha\mu}{}^{\mu} + \alpha_1\beta_2\epsilon^{\kappa\tau\mu}{}_{\alpha}F_{\kappa\tau,\mu} = 0\,, \tag{87}$$

$$2i\alpha_2\beta_4 T^{\mu}{}_{\mu\alpha} - \alpha_3\beta_7\epsilon^{\kappa\tau\mu}{}_{\alpha}\widetilde{T}_{\kappa,\tau\mu} = 0\,, \tag{88}$$

$$2i\alpha_3\beta_7 \widetilde{T}^{\mu}{}_{\mu\alpha} - \alpha_2\beta_4\epsilon^{\kappa\tau\mu}{}_{\alpha}T_{\kappa,\tau\mu} = 0\,, \tag{89}$$

$$i\epsilon^{\mu\nu\kappa\tau}\left[\alpha_2\beta_6\widetilde{R}_{\kappa\tau,\mu\nu} + \alpha_3\beta_8\widetilde{D}_{\kappa\tau,\mu\nu}\right] + 2\alpha_2\beta_5 R^{\mu\nu}{}_{\mu\nu} + 2\alpha_3\beta_9 D^{\mu\nu}{}_{\mu\nu} = 0\,, \tag{90}$$

$$i\epsilon^{\mu\nu\kappa\tau}\left[\alpha_2\beta_5 R_{\kappa\tau,\mu\nu} + \alpha_3\beta_9 D_{\kappa\tau,\mu\nu}\right] + 2\alpha_2\beta_6\widetilde{R}^{\mu\nu}{}_{\mu\nu} + 2\alpha_3\beta_8\widetilde{D}^{\mu\nu}{}_{\mu\nu} = 0\,, \tag{91}$$

$$2i\alpha_2\beta_5 R_{\beta\mu}{}^{\mu\alpha} + 2i\alpha_3\beta_9 D_{\beta\mu}{}^{\mu\alpha} + \alpha_2\beta_6\epsilon^{\nu\alpha}{}_{\lambda\beta}\widetilde{R}^{\lambda\mu}{}_{\mu\nu} + \alpha_3\beta_8\epsilon^{\nu\alpha}{}_{\lambda\beta}\widetilde{D}^{\lambda\mu}{}_{\mu\nu} = 0\,, \tag{92}$$

$$2i\alpha_1\beta_2 F^{\lambda\mu}{}_{\mu} - 2i\alpha_2\beta_4 T_{\mu}{}^{\mu\lambda} + \alpha_1\beta_3\epsilon^{\kappa\tau\mu\lambda}\widetilde{F}_{\kappa\tau,\mu} + \alpha_3\beta_7\epsilon^{\kappa\tau\mu\lambda}\widetilde{T}_{\kappa,\tau\mu} = 0\,, \tag{93}$$

$$2i\alpha_1\beta_3 \widetilde{F}^{\lambda\mu}{}_{\mu} - 2i\alpha_3\beta_7 \widetilde{T}_{\mu}{}^{\mu\lambda} + \alpha_1\beta_2\epsilon^{\kappa\tau\mu\lambda}F_{\kappa\tau,\mu} + \alpha_2\beta_4\epsilon^{\kappa\tau\mu\lambda}T_{\kappa,\tau\mu} = 0\,, \tag{94}$$

$$\begin{aligned}
&\alpha_1\beta_1(2G^{\lambda}{}_{\alpha} - g^{\lambda}{}_{\alpha}G^{\mu}{}_{\mu}) - 2\alpha_2\beta_5(2R^{\lambda\mu}{}_{\mu\alpha} + 2R_{\alpha\mu}{}^{\mu\lambda} + g^{\lambda}{}_{\alpha}R^{\mu\nu}{}_{\mu\nu}) + \\
+\ &2\alpha_3\beta_9(2D^{\lambda\mu}{}_{\mu\alpha} + 2D_{\alpha\mu}{}^{\mu\lambda} + g^{\lambda}{}_{\alpha}D^{\mu\nu}{}_{\mu\nu}) + 2i\alpha_3\beta_8(\epsilon_{\kappa\alpha}{}^{\mu\nu}\widetilde{D}^{\kappa\lambda}{}_{\mu\nu} - \epsilon^{\kappa\tau\mu\lambda}\widetilde{D}_{\kappa\tau,\mu\alpha}) - \\
-\ &2i\alpha_2\beta_6(\epsilon_{\kappa\alpha}{}^{\mu\nu}\widetilde{R}^{\kappa\lambda}{}_{\mu\nu} - \epsilon^{\kappa\tau\mu\lambda}\widetilde{R}_{\kappa\tau,\mu\alpha}) = 0\,,
\end{aligned} \tag{95}$$

$$\begin{aligned}
&2\alpha_3\beta_8(2\widetilde{D}^{\lambda\mu}{}_{\mu\alpha} + 2\widetilde{D}_{\alpha\mu}{}^{\mu\lambda} + g^{\lambda}{}_{\alpha}\widetilde{D}^{\mu\nu}{}_{\mu\nu}) - 2\alpha_2\beta_6(2\widetilde{R}^{\lambda\mu}{}_{\mu\alpha} + 2\widetilde{R}_{\alpha\mu}{}^{\mu\lambda} + \\
+\ &g^{\lambda}{}_{\alpha}\widetilde{R}^{\mu\nu}{}_{\mu\nu}) + +2i\alpha_3\beta_9(\epsilon_{\kappa\alpha}{}^{\mu\nu}D^{\kappa\lambda}{}_{\mu\nu} - \epsilon^{\kappa\tau\mu\lambda}D_{\kappa\tau,\mu\alpha}) - \\
-\ &2i\alpha_2\beta_5(\epsilon_{\kappa\alpha}{}^{\mu\nu}R^{\kappa\lambda}{}_{\mu\nu} - \epsilon^{\kappa\tau\mu\lambda}R_{\kappa\tau,\mu\alpha}) = 0\,,
\end{aligned} \tag{96}$$

$$\begin{aligned}
&\alpha_1\beta_2(F^{\alpha\beta,\lambda} - 2F^{\beta\lambda,\alpha} + F^{\beta\mu}{}_{\mu}g^{\lambda\alpha} - F^{\alpha\mu}{}_{\mu}g^{\lambda\beta}) - \\
-\ &\alpha_2\beta_4(T^{\lambda,\alpha\beta} - 2T^{\beta,\lambda\alpha} + T_{\mu}{}^{\mu\alpha}g^{\lambda\beta} - T_{\mu}{}^{\mu\beta}g^{\lambda\alpha}) + \\
+\ &\frac{i}{2}\alpha_1\beta_3(\epsilon^{\kappa\tau\alpha\beta}\widetilde{F}_{\kappa\tau}{}^{\lambda} + 2\epsilon^{\lambda\kappa\alpha\beta}\widetilde{F}_{\kappa\mu}{}^{\mu} + 2\epsilon^{\mu\kappa\alpha\beta}\widetilde{F}^{\lambda}{}_{\kappa,\mu}) - \\
-\ &\frac{i}{2}\alpha_3\beta_7(\epsilon^{\mu\nu\alpha\beta}\widetilde{T}^{\lambda}{}_{\mu\nu} + 2\epsilon^{\nu\lambda\alpha\beta}\widetilde{T}^{\mu}{}_{\mu\nu} + 2\epsilon^{\mu\kappa\alpha\beta}\widetilde{T}_{\kappa,\mu}{}^{\lambda}) = 0\,.
\end{aligned} \tag{97}$$

They are the results of contractions of the field function (80) with six antisymmetric matrices, as above. Furthermore, one should recover the relations (70-75) in the particular case when $\alpha_3 = \beta_3 = \beta_6 = \beta_9 = 0$ and $\alpha_1 = \alpha_2 = \beta_1 = \beta_2 = \beta_4 = \beta_5 = \beta_7 = \beta_8 = 1$.

As a discussion, we note that in such a framework we have physical content because only certain combinations of field functions can be equal to zero. In general, the fields $F_{\kappa\tau}{}^{\mu}$, $\widetilde{F}_{\kappa\tau}{}^{\mu}$, $T_{\kappa}{}^{\mu\nu}$, $\widetilde{T}_{\kappa}{}^{\mu\nu}$, and $R_{\kappa\tau}{}^{\mu\nu}$, $\widetilde{R}_{\kappa\tau}{}^{\mu\nu}$, $D_{\kappa\tau}{}^{\mu\nu}$, $\widetilde{D}_{\kappa\tau}{}^{\mu\nu}$ can correspond to different physical states and the equations above describe couplings one state with another.

Furthermore, from the set of equations (81-84) one obtains the *second*-order equation for the symmetric traceless tensor of the second rank ($\alpha_1 \neq 0$, $\beta_1 \neq 0$):

$$\frac{1}{m^2}\left[\partial_\nu \partial^\mu G_{\kappa}{}^{\nu} - \partial_\nu \partial^\nu G_{\kappa}{}^{\mu}\right] = G_{\kappa}{}^{\mu}\,. \tag{98}$$

After the contraction in indices κ and μ this equation is reduced to

$$\partial_\mu G^{\mu}{}_{\alpha} = F_\alpha\,, \tag{99}$$

$$\frac{1}{m^2}\partial_\alpha F^\alpha = 0\,, \tag{100}$$

i. e., to the equations connecting the analogue of the energy-momentum tensor and the analogue of the 4-vector potential (the additional notoph field as opposed to the Logunov theory?). As we showed in our recent work [30] the longitudinal potential may have importance in the construction of electromagnetism (see also the works on the notoph and notivarg concept [31]). Moreover, according to the Weinberg theorem [9] for massless particles it is the gauge part of the 4-vector potential $\sim \partial_\mu \chi$, which is the physical field. The case, when the longitudinal potential is equated to zero, can be considered as a particular case only. This case may be relevant to some physical situation but hardly to be considered as a basis for unification. Further investigations may provide additional foundations to "surprising" similarities of gravitational and electromagnetic equations in the low-velocity limit, Refs. [32, 33, 34, 36].

5. Interactions with Fermions.

The possibility of terms as $\sigma \cdot [\mathbf{A} \times \mathbf{A}^*]$ appears to be related to the matters of chiral interactions [38, 39]. As we are now convinced, the Dirac field operator can be always presented as a superposition of the self- and anti-self charge conjugate field operators (cf. Ref. [37]). The anti-self charge conjugate part can give the self charge conjugate part after multiplying by the γ^5 matrix, and *vice versa*. We derived[14]

$$[i\gamma^\mu D^*_\mu - m]\psi_1^s = 0\,, \tag{102}$$

or[15]

$$[i\gamma^\mu D_\mu - m]\psi_2^a = 0\,. \tag{104}$$

Both equations lead to the terms of interaction such as $\sigma \cdot [\mathbf{A} \times \mathbf{A}^*]$ provided that the 4-vector potential is considered as a complex function(al). In fact, from (102) we have:

$$i\sigma^\mu \nabla_\mu \chi_1 - m\phi_1 = 0\,, \tag{105}$$

$$i\tilde{\sigma}^\mu \nabla^*_\mu \phi_1 - m\chi_1 = 0\,. \tag{106}$$

And, from (104) we have

$$i\sigma^\mu \nabla^*_\mu \chi_2 - m\phi_2 = 0\,, \tag{107}$$

$$i\tilde{\sigma}^\mu \nabla_\mu \phi_2 - m\chi_2 = 0\,. \tag{108}$$

[14] The anti-self charge conjugate field function ψ_2 can also be used. The equation has then the form:

$$[i\gamma^\mu D^*_\mu + m]\psi_2^a = 0\,. \tag{101}$$

[15] The self charge conjugate field function ψ_1 also can be used. The equation has the form:

$$[i\gamma^\mu D_\mu + m]\psi_1^s = 0\,. \tag{103}$$

As readily seen, in the cases of alternative choices we have opposite charges in the terms of the type $\sigma \cdot [\mathbf{A} \times \mathbf{A}^*]$ and in the mass terms.

The meanings of σ^μ and $\tilde{\sigma}^\mu$ are obvious from the definition of γ matrices. The derivatives are defined:

$$D_\mu = \partial_\mu - ie\gamma^5 C_\mu + eB_\mu\,, \quad \nabla_\mu = \partial_\mu - ieA_\mu\,, \tag{109}$$

and $A_\mu = C_\mu + iB_\mu$. Thus, relations with the magnetic monopoles can also be established.

From the above system we extract the terms as $\pm e^2\sigma^i\sigma^j A_i A_j^*$, which lead to the discussed terms [38, 39].[16] Furthermore, one can come to the same conclusions not applying to the constraints on the creation/annihilation operators (which we have previously chosen for clarity and simplicity in Ref. [39]). It is possible to work with self/anti-self charge conjugate fields and the Majorana *anzatzen.* Thus, in the considered cases it is the γ^5 transformation which distinguishes various field configurations (helicity, self/anti-self charge conjugate properties etc) in the coordinate representation.

6. Boson Interactions.

The most general relativistic-invariant Lagrangian for the symmetric 2nd-rank tensor is

$$\begin{aligned}\mathcal{L} = &-\alpha_1(\partial^\alpha G_{\alpha\lambda})(\partial_\beta G^{\beta\lambda}) - \alpha_2(\partial_\alpha G^{\beta\lambda})(\partial^\alpha G_{\beta\lambda}) \\ &- \alpha_3(\partial^\alpha G^{\beta\lambda})(\partial_\beta G_{\alpha\lambda}) + m^2 G_{\alpha\beta}G^{\alpha\beta}\,.\end{aligned} \tag{110}$$

It leads to the equation

$$\left[\alpha_2\partial^2 + m^2\right] G^{\{\mu\nu\}} + (\alpha_1+\alpha_3)\partial^{\{\mu|}(\partial_\alpha G^{\alpha|\nu\}}) = 0\,. \tag{111}$$

In the case $\alpha_2 = 1 > 0$ and $\alpha_1 + \alpha_3 = -1$ it coincides with Eq. (98). There is no any problem to obtain the dynamical invariants for the fields of the spin 2 from the above Lagrangian. The mass dimension of $G^{\mu\nu}$ is $[energy]^1$.

We now present possible relativistic interactions of the symemtric 2nd-rank tensor. They should be the following ones:

$$\mathcal{L}^{int}_{(1)} \sim G_{\mu\nu}F^\mu F^\nu\,, \tag{112}$$

$$\mathcal{L}^{int}_{(2)} \sim (\partial^\mu G_{\mu\nu})F^\nu\,, \tag{113}$$

$$\mathcal{L}^{int}_{(3)} \sim G_{\mu\nu}(\partial^\mu F^\nu)\,. \tag{114}$$

The term $\sim (\partial_\mu G^\alpha_{\ \alpha})F^\mu$ vanishes due to the constraint of tracelessness. Obviously, these interactions cannot be obtained from the free Lagrangian (110) just by the covariantization of the derivative $\partial_\mu \to \partial_\mu + gF_\mu$.

It is also interesting to note that thanks to the possible terms

$$V(F) = \beta_1(F_\mu F^\mu) + \beta_2(F_\mu F^\mu)(F_\nu F^\nu) \tag{115}$$

we can give the mass to the G_{00} component of the spin-2 field. This is due to the possibility of the Higgs spontaneous symmetry breaking [40]

$$F^\mu(x) = \begin{pmatrix} v + \partial_0\chi(x) \\ g^1 \\ g^2 \\ g^3 \end{pmatrix}, \tag{116}$$

[16] I am grateful to Prof. S. Esposito for the e-mail communications (1997-98) on the alternative proof of the considered interaction. We would like to note that the terms of the type $\sigma \cdot [\mathbf{A} \times \mathbf{A}^*]$ can be reduced to $(\sigma \cdot \nabla)V$, where V is the scalar potential.

with v being the vacuum expectation value, $v^2 = (F_\mu F^\mu) = -\beta_1/2\beta_2 > 0$. Other degrees of freedom of the 4-vector field are removed since they can be interpreted as the Goldstone bosons. It was stated that "for any continuous symmetry which does not preserve the ground state, there is a massless degree of freedom which decouples at low energies. This mode is called the Goldstone (or Nambu-Goldstone) particle for the symmetry". As usual, the Higgs mechanism and the Goldstone modes should be important in giving masses to the three vector bosons.[17] As one can easily see, this expression does not permit an arbitrary phase for F^μ, which is possible only if the 4-vector would be the complex one.

Next, due to the Lagrangian interaction of fermions with notoph are of the order e^2 since the beginning (as opposed to the interaction with the 4-vector potential A_μ), it is more difficult to observe it. However, as far as I know the theoretical precision calculus in QED (the Landé factor, the anomalous magnetic moment, the hyperfine splittings in positronium and muonium, and the decay rate of o-Ps and p-Ps) are about the order corresponding to the 4th-5th loops, where the difference may appear with the experiments [41, 42].

7. Conclusions.

We considered the Bargmann-Wigner formalism to derive the equations for the AST field and for the symmetric tensor of the 2nd rank. We introduced additional scalar normalization field in the Bargmann-Wigner formalism in order to take into account possible physical significance of the Ogievetskiĭ-Polubarinov–Kalb-Ramond modes. We introduced the additional symmetric matrix in the Bargmann-Wigner expansion ($\gamma^5\sigma^{\mu\nu}R$) in order to take into account the dual fields. The consideration is similar to Ref. [43].

Furthermore, we discussed the interactions of notoph, photon and graviton (and, probably, notivarg[18]). For instance, the interaction notoph-graviton may give the mass to spin-2 particles in the way which is similar to the spontaneous-symmetry-breaking Higgs formalism.

Acknowledgements

I am grateful to the referee of "International Journal of Modern Physics" and "Foundation of Physics", whose advice of the mass dimensions (normalizations) of the fields was very useful. I acknowledge discussions with participants of recent conferences on Symmetries. I am thankful to organizers and *El Colegio Nacional* for partial financial support during the QTS-8.

References

[1] V. V. Dvoeglazov, Helv. Phys. Acta **70** (1997) 677, ibid. 686, ibid. 697.
[2] V. V. Dvoeglazov, Int. J. Theor. Phys. **37** (1998), 1915; hep-th/9611068.
[3] V. V. Dvoeglazov, Electromagnetic Phenomena **1**, No. 4 (1998) 465; hep-th/9604148.
[4] V. V. Dvoeglazov, Found. Phys. Lett. **10** (1997) 383.
[5] V. V. Dvoeglazov, Czech. J. Phys. **50** (2000) 225; , hep-th/9712036.
[6] D. V. Ahluwalia and D. J. Ernst, Mod. Phys. Lett. A**7** (1992) 1967.
[7] D. V. Ahluwalia and D. J. Ernst, Int. J. Mod. Phys. E**2** (1993) 397; D. V. Ahluwalia, M. B. Johnson and T. Goldman, Phys. Lett. B**316** (1993) 102.
[8] D. V. Ahluwalia and M. Sawicki, Phys. Rev. D**47** (1993) 5161.

[17] It is interesting to note the following statement (given without references in wikipedia.org): "In general, the phonon is effectively the Nambu-Goldstone boson for spontaneously broken Galilean/Lorentz symmetry. However, in contrast to the case of internal symmetry breaking, when spacetime symmetries are broken, the order parameter need not be a scalar field, but may be a tensor field, and the corresponding independent massless modes may now be fewer than the number of spontaneously broken generators, because the Goldstone modes may now be linearly dependent among themselves: e.g., the Goldstone modes for some generators might be expressed as gradients of Goldstone modes for other broken generators."

[18] In order to analize its dynamical invariants and interactions one should construct the Lagrangian from the analogs of the Riemann tensor $\widetilde{D}^{\mu\nu,\alpha\beta}$.

[9] S. Weinberg, Phys. Rev. B**133** (1964) 1318; ibid. B**134** 882.
[10] R. H. Tucker and C. L. Hammer, Phys. Rev. D**3** (1971) 2448.
[11] E. Durandin and A. Erschow, Phys. Z. Sowjet. **12** (1937) 466; F. Belinfante, Physica **6** (1939) 887.
[12] V. I. Ogievetskiĭ and I. V. Polubarinov, Sov. J. Nucl. Phys. **4** (1967) 156.
[13] K. Hayashi, Phys. Lett. B**44** (1973) 497.
[14] M. Kalb and P. Ramond, Phys. Rev. D**9** (1974) 2273.
[15] H. C. Ohanian, Am. J. Phys. **54** (1986) 500.
[16] L. V. Avdeev and M. V. Chizhov, Phys. Lett. B**321** (1994) 212; hep-th/9407067; M. V. Chizhov, Phys. Part. Nucl. **42** (2011) 93.
[17] D. Luriè, *Particles and Fields.* (Interscience Publishers, 1968), Chapter 1.
[18] N. N. Bogoliubov and D. V. Shirkov, *Introduction to the Theory of Quantized Fields.* (Moscow, Nauka, 1973).
[19] Yu. V. Novozhilov, *Introduction to Elementary Particle Theory.* (Pergamon Press, Oxford, 1975).
[20] C. Itzykson and J.-B. Zuber, *Quantum Field Theory.* (McGraw-Hill Book Co. New York, 1980).
[21] S. Weinberg, *The Quantum Theory of Fields. Vol. I. Foundations.* (Cambridge University Press, 1995), Chapter 5.
[22] V. Bargmann and E. P. Wigner, Proc. Natl. Acad. Sci. (USA) **34** (1948) 211.
[23] D. Han, Y. S. Kim and D. Son, Phys. Lett. **131**B (1983) 327; Y. S. Kim, Int. J. Mod. Phys. A**12** (1997) 71.
[24] V. V. Dvoeglazov, Hadronic J. Suppl. **12** (1997) 221.
[25] M. Kirchbach, Mod. Phys. Lett. A**12** (1997) 2373.
[26] N. D. Sen Gupta, Nucl. Phys. B**4** (1967) 147; A. Raspini, Fizika B**5** (1996) 159.
[27] A. C. Tam and W. Happer, Phys. Rev. Lett. **38** (1977) 278.
[28] P. C. Naik and T. Pradhan, J. Phys. A**14** (1981) 2795; T. Pradhan, R. P. Malik and P. C. Naik, Pramana J. Phys. **24** (1985) 77.
[29] G. Marques and D. Spehler, Mod. Phys. Lett. A**13** (1998) 553.
[30] V. V. Dvoeglazov, Phys. Scr. **64** (2001) 201, physics/9804010.
[31] S. Deser, Nucl. Phys. B**184** (1981) 333; J. Rembieliński and W. Tybor, Acta Phys. Polon. B**22** (1991) 439; ibid. 447; M. Bakalarska and W. Tybor, AIP Conf. Proc. **453** (1998) 312, hep-th/9801065; idem., *On the Deser-Siegel-Townsend Notivarg.* Preprint hep-ph/9801216, Łódź, Poland .
[32] S. Weinberg, *Gravitation and Cosmology.* (John Wiley & Sons, New York, 1972).
[33] O. D. Jefimenko, *Causality, Electromagnetic Induction and Gravitation.* (Electret Sci. Co., Star City, 1992); T. Chang, Galilean Electrodynamics **3** (1992) 36.
[34] A. Logunov and M. Mestvirishvili, *The Relativistic Theory of Gravitation.* (Mir Publishers, Moscow, 1989).
[35] V. L. Ginzburg, D. A. Kirzhnitz, public discussions about the RTG (some of their seminars have been attended by me since 1983).
[36] E. A. Notte-Cuello and W. A. Rodrigues Jr., Int. J. Mod. Phys. D**16** (2007) 1027, math-ph/0608017.
[37] A. Barut and G. Ziino, Mod. Phys. Lett. A**8** (1993) 1099; G. Ziino, Int. J. Mod. Phys. A**11** (1996) 2081.
[38] V. V. Dvoeglazov, Electromagnetic Phenomena, **2**, No. 1 (5) (2001) 3; hep-ph/9801287.
[39] V. V. Dvoeglazov, Acta Phys. Slovaca **50** (2000) 629.
[40] P. Higgs, Phys. Rev. Lett. **13** (1964) 508.
[41] V. V. Dvoeglazov, R. N. Faustov and Yu. N. Tyukhtyaev, Phys. Part. Nucl. **25** (1994) 58.
[42] T. Kinoshita, http://www.riken.jp/lab-www/theory/colloquium/kinoshita.pdf .
[43] V. V. Dvoeglazov, Int. J. Mod. Phys. CS **3** (2011) 121.

Journal of Physics: Conference Series **545** (2014) 012005 doi:10.1088/1742-6596/545/1/012005

Noncommutative Quantum Anisotropic cosmology in K-essence

Abraham Espinoza-García, J. Socorro

Departamento de Física de la DCeI de la Universidad de Guanajuato-Campus León
C.P. 37150, Guanajuato, México

E-mail: abraham@fisica.ugto.mx

E-mail: socorro@fisica.ugto.mx

Abstract. We study a canonical noncommutative extension of the Bianchi type I Minisuperspace, with a barotropic perfect fluid, in the context of a simplified form of k-essence known as the Sáez-Ballester theory. Noncommutativity is implemented in the pure gravitational sector. The corresponding noncommutative Wheeler-DeWitt equation is constructed and solved. Quantum solutions are obtained for any value of the barotropic parameter. It is seen that the effect of this particular noncommutativity is that of modulating the amplitude of the commutative wave function.

1. Introduction

In the early 2000's Armendariz et al proposed a minimally coupled scalar field with non standard kinetic term as a dynamical candidate for dark energy which is known today as K-essence [1] (as opposed to Quintessence which models dark energy through the potential term of a minimally coupled standard scalar field). K-essence is based on the idea of a dynamical attractor solution which causes it to act as a cosmological constant only at the onset of matter domination. More recently, attempts to unify the description of inflation, dark energy and dark matter through one or several k-essence fields have been conducted [2].

Quite ago, in the 1980's, Saez and Ballester [3] formulated a scalar-tensor theory of gravitation in which the metric is coupled with a dimensionless scalar field with non standard kinetic term, which suggested a possible way to solve the missing matter problem in non-flat FRW cosmologies. The Saez-Ballester (SB) formalism can be seen now as a very particular case of the K-essence paradigm applied to dark matter.

Usually K-essence models are restricted to a Lagrangian density of the form

$$\mathrm{S} = \int \mathrm{d}^4\mathrm{x}\,\sqrt{-\mathrm{g}}\,\left[\mathrm{f}(\phi)\,\mathcal{G}(\mathrm{X}) - \mathrm{V}(\phi)\right], \qquad (1)$$

where the canonical kinetic energy is given by $\mathcal{G}(\mathrm{X}) = \mathrm{X} = -\frac{1}{2}\nabla_\mu\phi\nabla^\mu\phi$. In particular, the SB formalism is obtained in the case in which the potential term is null and $\mathcal{G}(\mathrm{X}) = \mathrm{X}$.

The quantized version of the K-essence paradigm, and in particular the SB formalism, has not yet been constructed, perhaps due to the difficulties in obtaining the ADM formalism for it. We circumvent this issue by reinterpreting the K-essence field as an exotic part of the matter

Journal of Physics: Conference Series **545** (2014) 012005 doi:10.1088/1742-6596/545/1/012005

sector (obtaining it from an energy-momentum tensor) and thus reformulating the theory as General Relativity. This in turn enables us to implement the ADM formulation [4] and thus achieve the corresponding Wheeler-DeWitt quantization of the model.

One of the simplest K-essence Lagrangian density is

$$\mathcal{L}_{\rm geo} = \left(\frac{\rm R}{2} + {\rm f}(\phi)\mathcal{G}({\rm X}) \right), \tag{2}$$

here, R is the scalar curvature, and $f(\phi)$ is an arbitrary function of the scalar field and the potential for it is absent.

From the Lagrangian (2) we can build the complete action

$${\rm I} = \int_\Sigma \sqrt{-{\rm g}}(\mathcal{L}_{\rm geo} + \mathcal{L}_{\rm mat}){\rm d}^4{\rm x}, \tag{3}$$

where $\mathcal{L}_{mat}$ is the matter Lagrangian and g is the determinant of the metric tensor. The field equations for this theory are

$$\begin{aligned} {\rm G}_{\alpha\beta} + {\rm f}(\phi)\left[\mathcal{G}_{\rm X}\phi_{,\alpha}\phi_{,\beta} + \mathcal{G}{\rm g}_{\alpha\beta}\right] &= {\rm T}_{\alpha\beta}, & (4) \\ {\rm f}(\phi)\left[\mathcal{G}_{\rm X}\phi^{;\beta}_{;\beta} + \mathcal{G}_{\rm XX}{\rm X}_{;\beta}\phi^{;\beta}\right] + \frac{\rm df}{{\rm d}\phi}\left[\mathcal{G} - 2{\rm X}\mathcal{G}_{\rm X}\right] &= 0, & (5) \end{aligned}$$

where we work in units such that $8\pi G = 1$ and, as usual, the semicolon means a covariant derivative while G_X denotes differentiation of $G(X)$ with respect to X.

The same set of equations(4,5) is obtained if we consider the scalar field $X(\phi)$ as part of the matter content, i.e. $\mathcal{L}_{\rm X,\phi} = {\rm f}(\phi)\mathcal{G}({\rm X})$ with the corresponding energy-momentum tensor

$$\mathcal{T}_{\alpha\beta} = {\rm f}(\phi)\left[\mathcal{G}_{\rm X}\phi_{,\alpha}\phi_{,\beta} + \mathcal{G}({\rm X}){\rm g}_{\alpha\beta}\right].. \tag{6}$$

Making an analogy with a perfect fluid, we notice that the case $\mathcal{G}(X) = X$ corresponds to a stiff fluid.

On the other hand, the idea of Noncommutativity in spacetime is quite old (1947) [5]. It was proposed as an attempt to regularize Quantum Field Theory before the renormalization program was established. Due to its non-local behavior, Noncommutativity was quickly forgotten after renormalization proved to be successful. In the 1970's M. Flato and co-workers proposed an alternative path to quantization [6], in which a deformation of the Poisson structure of classical phase space is performed and is encoded in the moyal $\star$-product [7] and generalizations of it. In the early 1980's mathematicians led by A. Connes succeeded in formulating what they called Noncommutative Geometry [8], motivated by generalizating a classic theorem characterizing C^*-algebras.

Recently, the noncommutative paradigm has resurrected, mainly due to results in String Theory [9], in which Yang-Mills theories in a noncommutative space arise in different circumstances as effective theories when taking certain simple limits (for instance, the low energy limit). This renewed interest has led to a deeper understanding, from the physical and mathematical points of view, of noncommutative field theory (for a review see [10]). It is believed that in a full quantum theory of gravity the continuum picture of spacetime would no longer be consistent at distances comparable to the Planck length $\ell_p \sim 10^{-35}$cm, a quantization of spacetime itself could be in order. Furtheremore, since quite ago, it has been argued [11] that measurements of position can not be performed to better accuracies than the Planck length, since spacetime itself would be modified due to the energy required to peform such measurments.

Journal of Physics: Conference Series **545** (2014) 012005 doi:10.1088/1742-6596/545/1/012005

A possible way to model this effects could be via an uncertainty relation for the spacetime coordinates of the form

$$[x^i, x^j] = i\theta^{ij} \tag{7}$$

This commutation relation is the starting point of noncommutative field theory. Attempts to implement this idea have led to different proposals for a noncommutative theory of gravity [12]. Different incarnations of noncommutative gravity have a common feature, they are highly non-linear theories, which makes them very difficult to work with.

In the study of cosmological models spacetime is highly symmetric, as a consequence, the gravitational field can be described in terms of a finite number of degrees of freedom. We are not dealing with a field theory anymore. A possible way to study noncommutativity effects in the early universe was proposed by Garcia-Compean et al [13]. They implemented noncommutativity in configuration space (in contrast to noncommutative spacetime), but only after a symmetry reduction of spacetime had been imposed and the Wheeler-DeWitt quantization had been carried, giving rise to a noncommutative quantum cosmology (mathematically similar to noncommutative quantum mechanics [14]). The idea is that perhaps this effective noncommutativity could incorporate novel effects and insights of a full quantum theory of the gravitational field, alongside with providing a simple framework for studying the implications of possible noncommutativity effects in the (early) universe. Some works along this line have been conducted. For instance, the noncommutativity of the Friedmann-Robertson-Walker cosmology has been studied, as well as some of the Bianchi Class A models [15].

Since when gravity is coupled to matter the purely gravitational sector of minisuperspace carries information about pure spacetime, and it is the spacetime noncommutativity effects which in principle one wants to approximate in an effective way, the natural step would be to incorporate noncommutativity only in the gravitational degrees of freedom of the resulting coupled minisuperspace. In the present work we implement this idea for the Bianchi Type I model, with the coupling of a particularly simple K-essence field and orindary matter being modelled by a perfect barotropic fluid.

The manuscript is organized as follows: In Section II we focus on recalling the emergence of the moyal $\star$-product within the Weyl quantization prescription and how it can be applied to Quantum Mechanics. Section III deals with the implementation of noncommutativity via the moyal $\star$ product in the Minisuperspace of the Bianchi type I at the quantum level. Noncommutative quantum general solutions are found by solving the corresponding noncommutative Wheeler-DeWitt equation. Section IV is devoted to point out some conclusions and perspectives of the work.

2. Deformations in Quantum Mechanics

A noncommutativity in classical and quantum phase space can be achieved by deforming the flat Poisson structure in those spaces. One can encode this classical and quantum phase space noncommutativity in a deformation of the product, the so called $\star$-product. This proceedure gives rise to Noncommutative Classical Mechanics and Noncommutative Quantum Mechanics. It also provides a third road to quantization, the deformation quantization program. A noncommutative spacetime can also be encoded with the help of a $\star$-product, giving rise to the study of field theory in noncommutative spacetimes.

2.1. Weyl-Wigner-Moyal Correspondence

Noncommutativity can be introduced through a $\star$-product. As stated above, this kind of noncommutativity has been developed in connection with deformation quantization. A $\star$-product is an asociative noncommutative product which encodes the noncommutativity but still permits us to work in the usual corresponding commutative space (phase space, spacetime, etc.). The Weyl-Wigner-Moyal (WWM) correspondence relates a phase space function f with its operator counterpart $\mathcal{W}(f) = \hat{f}$, it is based on the following map [17]

$$\hat{f}(\hat{q},\hat{p}) = \int \tilde{f}(\xi,\eta) e^{\frac{i}{\hbar}(\hat{p}_a\xi^a + \hat{q}^a\eta_a)} w(\xi,\eta) d^l\xi d^l\eta \tag{8}$$

where $\tilde{f}$ is the Fourier transform of f, w a weight function, and $dim(T^*Q) = 2l$; with $\hat{q}^a, \hat{p}_b$ satisfying the canonical Heisenberg algebra,

$$[\hat{q}^a, \hat{p}_b] = i\hbar\delta^a_b; \quad [\hat{q}^a, \hat{q}^b] = 0 = [\hat{p}_a, \hat{p}_b] \tag{9}$$

We define the $\star$-product with the help of this correspondence by

$$\mathcal{W}(f \star g) = \mathcal{W}(f) \cdot \mathcal{W}(g) \tag{10}$$

that is, $f \star g$ is the Weyl symbol of $\hat{f}\hat{g}$. For this case of the Heisenberg algebra (9) and in the symmetric ordering ($w = 1$) we have

$$f \star g = \exp\left(\frac{i\hbar}{2} P(f,g)\right) = f(x)\exp\left(\frac{i\hbar}{2}\overleftarrow{\partial}_\mu \omega^{\mu\nu}\overrightarrow{\partial}_\nu\right) g(x) = fg + \frac{1}{2}\{f,g\} + \mathcal{O}(\hbar^2) \tag{11}$$

where $P^r(f,g) = \omega^{\mu_1\nu_1}\cdots\omega^{\mu_r\nu_r}(\partial_{\mu_1\ldots\mu_r}f)(\partial_{\nu_1\ldots\nu_r}g)$ is the rth power of the Poisson bracket bidifferential operator, with $\omega^{\mu\nu}$ the components of the usual flat Poisson structure; x^μ are the phase space coordinates, denoted collectevely as x, the first n being the configuration coordinates q^a, the second n being the momenta p_a. This is the Moyal $\star$-product [7], it replaces the ordinary pointwise multiplication in the algebra of functions defined in phase space. The Moyal $\star$-bracket $f \star g - g \star f$ is thus responsible for realize the canonical Heisenberg algebra (9). Hence, (11) encodes a deformation of the classical phase space which yields the canonical Heisenberg algebra. This product is the cornerstone of deformation quantization [6].

Now, consider the deformation of classical phase space which yields the algebra

$$[\hat{q}^a, \hat{p}_b] = i\hbar\delta^a_b; \quad [\hat{q}^a, \hat{q}^b] = i\theta\theta^{ab}; \quad [\hat{p}_a, \hat{p}_b] = i\kappa\kappa_{ab} \tag{12}$$

with θ^{ab}, κ_{ab} antisymmetric constat real matrices.

In light of the above prescription, we could encode this deformed Heisenberg algebra in a $\star$-product completely analogous to (11). The difference lies in the $\omega^{\mu\nu}$, which in this case would be of the form

$$i\begin{pmatrix} \theta\Theta & \hbar I \\ -\hbar I & \kappa \mathrm{K} \end{pmatrix} \tag{13}$$

This deformation quantization would lead us to a Noncommutative Quantum Mechanics, rather than the ordinary Quantum Mechanis obtained above. Since we do not view Quantum Mechanics as a deformation quantization of classical phase space, we will follow a different route.

2.2. Noncommutative Quantum Mechanics

We implement the $\star$-product at the quantum level (f, g in (11 are now defined on quantum phase space). We take the canonical Heisenberg algebra (9) as given, then we deform it to (12). The $\omega^{\mu\nu}$ is in this case of the form

$$i\begin{pmatrix} \theta\Theta & 0 \\ 0 & \kappa \mathrm{K} \end{pmatrix} \tag{14}$$

Here, the $\star$-bracket $\hat{f}\star\hat{g} - \hat{g}\star\hat{f}$ reduces to the ordinary bracket as the noncommutative parameters θ, κ go to zero (and thus the deformed algebra reduces to the usual quantum Heisenberg algebra), in contrast with the $\star$-bracket that arises in the context of deformation quantization. There, the canonical Heisenberg algebra (9), or a deformation of it like (12), is realized due to the Poisson bracket in the leading term in the deformation parameter.

The time evolution for a state ket $|\psi\rangle$ in abstract Hilbert space is given by the equation

$$i\hbar\frac{\partial}{\partial t}|\psi\rangle = \hat{H}|\psi\rangle \tag{15}$$

Now, in the case in which the noncommutativity enters only in the configuration sector of quantum phase space ($\kappa = 0$ in (14)), in the position representation ($\langle\mathbf{q}|\hat{\mathbf{q}}|\psi\rangle = \mathbf{q}\langle\mathbf{q}|\psi\rangle$, $\langle\mathbf{q}|\exp\left(\frac{i}{\hbar}\mathbf{a}\cdot\hat{\mathbf{p}}\right)|\psi\rangle = \langle\mathbf{q}+\mathbf{a}|\psi\rangle$; with the abstract Hilbert space realized by the function space $L^2(\mathbb{R}^n, d^nq)$) and for the simple case in which $\hat{H} = \frac{1}{2m}\hat{\mathbf{p}}^2 + \hat{V}(\hat{\mathbf{q}})$, Schrödinger's equation for the wave function $\psi(\mathbf{q}) = \langle\mathbf{q}|\psi\rangle$ takes the form (the momenta and $\frac{\partial}{\partial t}$ terms remain the same under the $\star$ operation)

$$i\hbar\frac{\partial}{\partial t}\psi(\mathbf{q}) = -\frac{\hbar}{2m}\sum_{a=1}^{n}\partial_a^2\psi + V(\mathbf{q})\star\psi \tag{16}$$

so that the information about the deformation enters only through the potential term. The $\star$ operation has the effect of shifting the coordinates in the following way [14]

$$V(q^1,...,q^n)\star\psi = V(q^1 - \frac{1}{2}\theta^{1b}p_b, ..., q^n - \frac{1}{2}\theta^{nb}p_b)\psi \tag{17}$$

This shifting enables one to encode the noncommutativity by performing the transformation $q^{a\prime} = q - \theta^{ab}\frac{p_b}{2}$ in Schrödinger's wave equation equation (16) but with the ordinary product, since the new coordinates $\hat{q}'^a$ satisfy the algebra (12) with $\kappa = 0$. This is the kind of Noncommutative Quantum Mechanics that has been mostly studied [14].

3. Deformation of the Bianchi Type I Minisuperspace

3.1. Canonical Formulation

Let us recall here the canonical formulation in the ADM formalism of the diagonal Bianchi Class A models. The metric has the form

$$ds^2 = -(Ndt)^2 + e^{2\Omega(t)}\left(e^{2\beta(t)}\right)_{ij}\omega^i\omega^j = -d\tau^2 + e^{2\Omega(t)}\left(e^{2\beta(t)}\right)_{ij}\omega^i\omega^j, \tag{18}$$

where $\Omega(t)$ is a scalar, N the lapse function and $\beta_{ij}(t)$ a $3x3$ diagonal matrix, $\beta_{\mathrm{ij}} = \mathrm{diag}(\beta_+ + \sqrt{3}\beta_-, \beta_+ - \sqrt{3}\beta_-, -2\beta_+)$, ω^i are one-forms that characterize each cosmological Bianchi type model and obey $d\omega^i = \frac{1}{2}C^i_{jk}\omega^j\wedge\omega^k$, C^i_{jk} the structure constants of the corresponding invariance group. For the Bianchi type I model we have

$$\omega^1 = dx^1 \qquad \omega^2 = dx^2, \qquad \omega^3 = dx^3$$

The total Lagrangian density then reads

$$\mathcal{L}_I = e^{3\Omega}\left[6\frac{\dot{\Omega}^2}{N} - 6\frac{\dot{\beta}_+^2}{N} - 6\frac{\dot{\beta}_-^2}{N} + \frac{f(\phi)}{N}\dot{\phi}^2 + 16\pi G N\rho\right], \tag{19}$$

using the standard definition of the momenta, $p_{q^\mu} = \frac{\partial\mathcal{L}}{\partial\dot{q}^\mu}$, where $q^\mu = (\Omega, \beta_+, \beta_-, \phi)$, we obtain

$$p_\Omega = \frac{12}{N}e^{3\Omega}\dot{\Omega}, \quad\rightarrow\quad \dot{\Omega} = \frac{N}{12}e^{-3\Omega}p_\Omega \tag{20}$$

$$p_+ = -\frac{12}{N}e^{3\Omega}\dot{\beta}_+, \quad\rightarrow\quad \dot{\beta}_+ = -\frac{N}{12}e^{-3\Omega}p_+ \tag{21}$$

$$p_- = -\frac{12}{N}e^{3\Omega}\dot{\beta}_-, \quad\rightarrow\quad \dot{\beta}_- = -\frac{N}{12}e^{-3\Omega}p_+ \tag{22}$$

$$p_\phi = \frac{2f}{N}e^{3\Omega}\dot{\phi}, \quad\rightarrow\quad \dot{\phi} = \frac{N}{2f}e^{-3\Omega}p_\phi \tag{23}$$

Journal of Physics: Conference Series **545** (2014) 012005 doi:10.1088/1742-6596/545/1/012005

and introducing them into the Lagrangian density, we obtain the canonical Lagrangian as $\mathcal{L}_{can} = p_{q^\mu}\dot{q}^\mu - N\mathcal{H}$,

$$\mathcal{L}_{can} = p_{q^\mu}\dot{q}^\mu - \frac{N}{24}e^{-3\Omega}\left\{p_\Omega^2 + \frac{6}{f(\phi)}p_\phi^2 - p_+^2 - p_-^2 - 384\pi G\mu_\gamma e^{3\Omega(1-\gamma)}\right\}, \quad (24)$$

so, the ADM action has the form,

$$I = \int\left(p_+ d\dot{\beta}_+ + p_- d\dot{\beta}_- + p_\Omega d\dot{\Omega} + p_\phi d\dot{\phi} + N\mathcal{H}dt\right), \quad (25)$$

where

$$\mathcal{H} = \frac{e^{-3\Omega}}{24}\left(-p_\Omega^2 - \frac{6}{f(\phi)}p_\phi^2 + p_+^2 + p_+^2 + m_\gamma e^{-3(\gamma-1)\Omega}\right), \quad (26)$$

with $m_\gamma = 384\pi G\mu_\gamma$.

3.2. Noncommutative Wheeler-DeWitt Equation

As stated in the introduction, it is believed that the notion of the spacetime continuum breaks down at scales in which the quantum nature of spacetime plays a mayor role. A possible scenario is a noncommutative spacetime, in which the notion of a point is replaced by that of a planck cell. We thus impose noncommutativity among the gravitational degrees of freedom only, since they carry information about spacetime itself.

We consider the deformation

$$[p_i, p_j] = 0, [q^i, p_j] = i\delta^i_j, [\Omega, \beta_+] = i\theta_1, [\Omega, \beta_-] = i\theta_2,$$
$$[\beta_+, \beta_-] = i\theta_3, [\Omega, \phi] = [\phi, \beta_+] = [\phi, \beta_-] = 0 \quad (27)$$

In view of (17), the noncommutative Wheeler-DeWitt (NCWDW) equation, corresponding to the deformed algebra (27), is therefore ($p_i \mapsto i\partial_i$)

$$\frac{e^{-3\Omega_{nc}}}{24}\left[\partial_\Omega^2 - \partial_+^2 - \partial_-^2 + \frac{6}{f(\phi)}\partial_\phi^2 + m_\gamma e^{-3(\gamma-1)\Omega_{nc}}\right]\Psi(\Omega, \beta_\pm, \phi) = 0 \quad (28)$$

where we have set $\mathcal{G}_X = 1$ and

$$\Omega_{nc} = \Omega - \tfrac{\theta_1}{2}p_+ - \tfrac{\theta_2}{2}p_- \quad (29)$$

Employing the semi-general factor ordering due to Hartle and Hawking [18], the NCWDW equation takes the form

$$\left[\partial_\Omega^2 - \partial_+^2 - \partial_-^2 + \frac{6}{f(\phi)}\partial_\phi^2 - \frac{6s}{f(\phi)\phi}\partial_\phi - q\partial_\Omega + m_\gamma e^{-3(\gamma-1)\Omega_{nc}}\right]\Psi = 0 \quad (30)$$

We further assume a solution of the form

$$\Psi(\Omega, \beta_\pm, \phi) = \mathcal{A}(\Omega)\mathcal{B}(\beta_\pm)\mathcal{C}(\phi) \quad (31)$$

where we take the ansatz $\mathcal{B} = e^{\pm ia\beta_+}e^{\pm ib\beta_-}$. Employing the Baker-Campbell-Hausdorff (BCH) relation

$$\exp^X \exp^Y = \exp\left(Y + [X, Y] + \frac{1}{2}[X, [X, Y]] + \frac{1}{3!}[X, [X, [X, Y]]] + ...\right) \quad (32)$$

we find

$$\frac{1}{\mathcal{A}}\frac{d^2\mathcal{A}}{d\Omega^2} - \frac{1}{\mathcal{B}}\frac{\partial^2\mathcal{B}}{\partial\beta_+^2} - \frac{1}{\mathcal{B}}\frac{\partial^2\mathcal{B}}{\partial\beta_-^2} + \frac{6}{f\mathcal{C}}\frac{\partial^2\mathcal{C}}{\partial\phi^2} - \frac{6s}{\phi f}\frac{1}{\mathcal{C}}\frac{\partial\mathcal{C}}{\partial\phi} - q\frac{1}{\mathcal{A}}\frac{\partial\mathcal{A}}{\partial\Omega} + m_\gamma e^{-3(\gamma-1)\Omega}e^{-\frac{3}{2}(\gamma-1)(a\theta_1+b\theta_2)} = 0 \quad (33)$$

Separating variables we obtain the following set of ordinary differential equations

$$\tfrac{d^2\mathcal{A}}{d\Omega^2} - q\tfrac{d\mathcal{A}}{d\Omega} + \left(\tilde{m}_\gamma e^{-3(\gamma-1)\Omega} - \alpha^2\right)\mathcal{A} = 0 \quad (34)$$

$$\tfrac{\partial^2\mathcal{B}}{\partial\beta_+^2} + \tfrac{\partial^2\mathcal{B}}{\partial\beta_-^2} - \kappa^2\mathcal{B} = 0 \quad (35)$$

$$\tfrac{d^2\mathcal{C}}{d\phi^2} + s\tfrac{d\mathcal{C}}{d\phi} + \tfrac{\beta^2 f}{6}\mathcal{C} = 0 \quad (36)$$

where $\kappa^2 = \alpha^2 + \beta^2$ and $\tilde{m}_\gamma = m_\gamma e^{-\frac{3}{2}(\gamma-1)(a\theta_1+b\theta_2)}$ carries information about the noncommutativity.

Now, equation (34) can be solved in the follwing way: perform the transformation

$$z = \frac{\sqrt{\tilde{m}_\gamma}}{\eta}e^{-\frac{3}{2}(\gamma-1)\Omega} \quad (37)$$

substituting in (34) and assuming a solution of the form

$$\mathcal{A}(\Omega) = z^{\zeta q}\mathcal{F}(z) \quad (38)$$

we get

$$\frac{9}{4}(\gamma-1)^2 z^{\zeta q}\left[z^2\frac{d^2\mathcal{F}}{dz^2} + z\left(1+2\zeta q+\frac{2}{3}\frac{q}{\gamma-1}\right)\frac{d\mathcal{F}}{dz} + \left(\frac{4\eta^2}{9(\gamma-1)^2}z^2 + \right.\right.$$
$$\left.\left. + q^2\left\{\zeta^2+\frac{2}{3}\frac{\zeta}{\gamma-1}\right\} - \frac{4\alpha^2}{9(\gamma-1)^2}\right)\mathcal{F}\right] = 0,$$

which can be written as

$$z^2\frac{d^2\mathcal{F}}{dz^2} + z\frac{d\mathcal{F}}{dz} + \left[z^2 - \frac{1}{9(\gamma-1)^2}\left(q^2+4\alpha^2\right)\right]\mathcal{F} = 0, \quad (39)$$

when η and ζ are fixed to

$$\zeta = -\frac{1}{3(\gamma-1)}, \qquad \eta = \frac{3}{2}|\gamma-1| \quad (40)$$

This is the Bessel differential equation for the function $\mathcal{F}$.

(i) The solution for $\mathcal{A}(\Omega)$ for the case $\gamma \neq 1$ is then

$$\mathcal{A}_\gamma = \left(\frac{2\sqrt{\tilde{m}_\gamma}}{3|\gamma-1|}e^{-\frac{3}{2}(\gamma-1)\Omega}\right)^{-\frac{q}{3(\gamma-1)}}\left(c_\gamma J_\nu\left(\frac{2\sqrt{\tilde{m}_\gamma}}{3|\gamma-1|}e^{-\frac{3}{2}(\gamma-1)\Omega}\right) + d_\gamma Y_\nu\left(\frac{2\sqrt{\tilde{m}_\gamma}}{3|\gamma-1|}e^{-\frac{3}{2}(\gamma-1)\Omega}\right)\right), \quad (41)$$

where $\nu = \pm\sqrt{\frac{1}{9(\gamma-1)^2}\left(q^2+4\alpha^2\right)}$ and J_ν and Y_ν are the ordinary Bessel functions of the first and second kind, respectively.

(ii) For the case $\gamma = 1$ equation (34) takes the form

$$\frac{d^2A_1}{d\Omega^2} - q\frac{dA_1}{d\Omega} + \alpha_1^2 A_1 = 0, \qquad \alpha_1^2 = \tilde{m}_1 - \alpha^2, \quad (42)$$

whose solution is

$$A_1 = A_1 e^{\frac{q+\sqrt{q^2-4\alpha_1^2}}{2}\Omega} + A_2 e^{\frac{q-\sqrt{q^2-4\alpha_1^2}}{2}\Omega} \quad (43)$$

Finally, to obtain a solution for equation (36) we need to fix $f(\phi)$, we take

$$f(\phi) = \omega e^{m\phi} \tag{44}$$

in the original proposal of Sáez-Ballester $f(\phi) = \omega\phi^m$. The solution for (36) is

$$\mathcal{C}(\phi) = e^{-\frac{s\phi}{2}}\left[c\Gamma\left(1-\nu\right)J_{-\nu}\left(\frac{\sqrt{\frac{2}{3}}\sqrt{\beta^2\omega e^{m\phi}}}{m}\right) + d\Gamma\left(1+\nu\right)J_{\nu}\left(\frac{\sqrt{\frac{2}{3}}\sqrt{\beta^2\omega e^{m\phi}}}{m}\right)\right] \tag{45}$$

where $\nu = \frac{s}{m}$ is the order of the Bessel functions and $\Gamma(z)$ is Euler's gamma function.

As expected, this noncommutative quantum solutions reduce to the commutative ones (some of them are reported in Ref. [19]) as the noncommutative parameters θ_i go to zero.

4. Final Remarks

As can be seen from the relations (41),(43) and Ref. [19], the noncommutativity does not modify the form of the solutions, the noncommutative parameters contribute only through a multiplicative constant in the term $\tilde{m}_\gamma = e^{-\frac{3}{2}(\gamma-1)(a\theta_1+b\theta_2)}$ which takes part in the z term and in the argument of the Bessel functions in (41). A similar situation ocurs regarding solution (43). Therefore, the effect of this particular noncommutativity is to modulate the amplitude of the original commutative wave function of the Bianchi type I universe. We recover the original commutative solution in the limiting case when the noncommutative parameter θ vanishes.

Acknowledgments

This work was partially supported by CONACYT 167335, 179881 grants. PROMEP grants UGTO-CA-3 and UAM-I-43. A.E.G. is supported by a CONACYT graduate fellowship. This work is part of the collaboration within the Instituto Avanzado de Cosmología and Red PROMEP: Gravitation and Mathematical Physics under project *Quantum aspects of gravity in cosmological models, phenomenology and geometry of space-time.* Many calculations where done by Symbolic Program REDUCE 3.8.

[1] C. Armendariz-Picon, V. Mukhanov and P.J. Steinbardt, *Phys. Lett.* **85**, 4438 (2000); *Phys. Rev. D* **63**, 103510 (2001).
[2] De-Santiago, Josue and Cervantes-Cota, Jorge L., *Phys. Rev. D*,**83**,063502, (2011).
[3] D. Saez and V.J. Ballester, *Phys. Lett. A* **113**, 467 (1986).
[4] M.P. Ryan, *Hamiltonian cosmology*, (Springer, Berlin, 1972).
[5] H.S. Snyder, *Phys. Rev.*,**71**,38, (1947). *Quantized space-time*
[6] F. Bayen, M. Flato, C. Fronsdal, A. Lichnerowicz and D. Sternheimer, *Lett. Math. Phys.* **1** (1977), 521530; F. Bayen, M. Flato, C. Fronsdal, A. Lichnerowicz and D. Sternheimer *Ann. Phys.* **111**, 61 (1978); F. Bayen, M. Flato, C. Fronsdal, A. Lichnerowicz and D. Sternheimer *Ann. Phys.* **111**, 111 (1978).
[7] J. Moyal, *Proc. Camb. Phil. Soc.* **45**, 99 (1949).
[8] A. Connes, *Noncommutative Geometry*, Academic Press, New York (1994)
[9] N. Seiberg and E. Witten, *J. High Energy Phys.* **9906**, 032 (1999) *String theory and noncommutative geometry*, [arXiv:hep-th/9908142].
[10] M. R. Douglas, N. A. Nekrasov, *Rev. Mod. Phys.* **73**, 977. *Noncommutative field theory*, [arXiv:hep-th/0106048]; R. J. Szabo, *Phys. Rept.* **378**, 207 (2003) *Quantum field theory on noncommutative spaces*, [arXiv:hep-th/0109162].
[11] B.S. DeWitt, *Gravitation: an Introduction to Current Research* (1962)
[12] J.W. Moffat, *Phys. Lett. B* **491** 345 (2000); A.H. Chamsedine, *Commun. Math. Phys.* **218**, 283 (2001); A.H. Chamsedine, *Commun. Math. Phys.* **218**, 283 (2001); H. Garcia-Compean, O. Obregon, C. Ramirez and M. Sabido, *Phys. Rev. D* **68**, 044015 (2003). *Noncommutative Self-dual gravity*, [arXiv:hep-th/0506157]; H. Garcia-Compean, O. Obregon, and C. Ramirez, *Rev. Mex. Fís. S* **53**, 75 (2007).
[13] H. Garcia-Compean, O. Obregon, and C. Ramirez, *Phys. Rev. Lett.*, **88**, 161301 (2002). *Noncommutative quantum cosmology*, [arXiv:hep-th/0107250].

[14] L. Mezincescu, *Star operation in Quantum Mechanics*, [arXiv:hep-th/0007046]; J. Gamboa, M. Loewe and J.C. Rojas, *Phys. Rev. D* **64** 067901 (2001); M. Chaichian, M.M. Sheikh-Jabbari and A. Tureanu, *Phys. Rev. Lett.* **86**, 2716 (2001);
[15] W. Guzman, M. Sabido and J. Socorro, *Phys. Rev. D* **76**, 08732 (2007); J. Socorro, Luis O. Pimentel, C. Ortiz, M. Aguero, *Int. J. Theor. Phys.* **48**, 3567 (2009);
[16] W. Guzman, M. Sabido and J. Socorro, *Phys. Lett. B* **697**, 271 (2011).
[17] H. Weyl, *The Theory of Groups and Quantum Mechanics* Dover, New York (1931).
[18] S.W. Hawking, Nucl. Phys. B239, 257 (1984).
[19] M. Sabido, J. Socorro, L. Arturo Urena-Lopez, *Fizika B* **19**, 177 (2010).

Journal of Physics: Conference Series **545** (2014) 012006 doi:10.1088/1742-6596/545/1/012006

A relativistic axisymmetric approach to the galactic rotation curves problem

A Herrera–Aguilar[1,2] and U Nucamendi[2]

[1] Departamento de Física, Universidad Autónoma Metropolitana Iztapalapa, San Rafael Atlixco 186, CP 09340, México D. F., México.

[2] Instituto de Física y Matemáticas, Universidad Michoacana de San Nicolás de Hidalgo, Edificio C–3, Ciudad Universitaria, CP 58040, Morelia, Michoacán, México.

E-mail: aherreraaguilar@xanum.uam.mx

E-mail: ulises@ifm.umich.mx

Abstract.

It is known that galactic potentials can be kinematically linked to the observed red/blue shifts of the corresponding galactic rotation curves under a minimal set of assumptions (see [1] and [2] for details): *i*) that emitted photons come to us from stable timelike circular geodesic orbits of stars in a static spherically symmetric gravitational field, and *ii*) that these photons propagate to us along null geodesics. This relation can be established without appealing at all to a concrete theory of gravitational interaction. This kinematical spherically symmetric approach to the galactic rotation curves problem can be generalized to the stationary axisymmetric realm, which is precisely the symmetry that spiral galaxies possess [3]. Here we review the relativistic results obtained in the latter work. Namely, by making use of the most general stationary axisymmetric metric, we consider stable circular orbits of stars that emit signals which travel to a distant observer along null geodesics and express the galactic red/blue shifts in terms of three arbitrary metric functions, clarifying the contribution of the rotation as well as the dragging of the gravitational field. This stationary axisymmetric approach distinguishes between red and blue shifts emitted by circularly orbiting receding and approaching stars, respectively, even when they are considered with respect to the center of a spiral galaxy, indicating the need of precise measurements in order to confront predictions with observations. We also point out the difficulties one encounters in the attempt of determining the metric functions from observations and list some potential strategies to overcome them.

1. Introduction

The galactic rotation curves provide a direct method of determining the gravitational field inside a spiral galaxy since they have been measured for a great amount of galaxies [4]–[9]. These curves are obtained by measuring the red/blue shifts of light emitted from stars and from the 21 cm radiation from neutral gas clouds. The observations show evidence that the red/blue shifts z, or equivalently, the tangential velocities of rotation v, remain constant or decay more slowly than the Keplerian behaviour ($v^2 \sim 1/r$) up to distances far beyond the luminous radius of these galaxies. By performing a naive Newtonian analysis of this effect, one deduces that the energy density of the galaxies decreases approximately as r^{-2}, and hence, the mass of these bodies should increase as $m(r) \approx r$. Since the observed luminous galactic components do not produce this growing behaviour, a question arises: what is the reason of such an effect? Nowadays there is a strong believe that dark matter is responsible for it, being the major bounded constituent of galaxies and galaxy clusters ($\sim 25\%$ of the total energy density of our Universe [10]). However, one can naturally ask whether this large unseen mass does not produce a relevant gravitational red shift. On the other hand, in [11] it was shown for a large and complete sample of spiral galaxies that their luminous regions consist of stellar discs embedded in universal dark halos of constant density independently of the galaxy properties; moreover, it was shown that the Dark Matter halos made by the most likely dark particle are inconsistent with actual observations. Finally, alternative approaches to this problem like modifications of Newtonian dynamics based on these observations have also been developed [12]–[16]. In this review we shall analyze the problem on the basis of what it is directly observed: the photons red/blue shifts. This approach enables us to keep track of the effect of the underlying made assumptions, and to be aware of when they are not longer valid.

There have been previous relativistic approaches to the galactic rotation curves problem that make use of stationary axisymmetric metrics within the framework of general relativity. For instance, Matos *et al.* [17] have shown that the expression for the tangential velocity of test objects which follow a circular stable geodesic orbit in the equatorial plane can be expressed as a function of arbitrary metric coefficients of an axisymmetric *static* space-time. On the other hand, within the comoving dust solution of Cooperstock and collaborators (see e.g. [18]–[23]) it is claimed that the known data from galactic rotation curves can be described by non-linear general relativistic effects in galactic dynamics with at most relatively little extra matter. A crucial issue within this approach is that even though fields and velocities are small in a galaxy, it is not consistent to describe galactic dynamics through a Newtonian approximation since there are non-linear contributions of non-negligible size coming from Einstein equations. However, these models have received criticisms in several directions that point out to unphysical features like the need of an additional exotic matter source in the galactic disc [24] or infinite mass at large distances [25], the presence of singularities when continuing the interior solution into a consistent exterior configuration [26], the inconsistency of the use of comoving frames with the condition of differential rotation in a galaxy [27], etc. An important observation upon this model was pointed out in [28], namely, the stationary axisymmetric metric of the comoving dust solution of [18] does not possess the most general form since it has only three arbitrary functions instead of four, moreover, their assumed Weyl gauge (W=r) is not consistent with the Einstein equations since this metric function is not harmonic and hence, it does not belong to the most general class of stationary axisymmetric solutions, the Lewis-Papapetrou class, a circumstance that might be connected with the problems of the model.

The aim of this review paper is to provide a stationary axisymmetric kinematical description of the galactic rotation curves problem since this is a more realistic symmetry compared to the spherical one when studying spiral galaxies: the most accepted composition of spiral galaxies indicates that its main mass constituent is concentrated in a thin disc with a central bulge which are surrounded by a spherical halo [29]. Thus, it is commonly accepted that the main aspects of the galactic dynamics can be approximately described by (rotating) thin galactic disc models. We make use of the most general stationary axisymmetric metric and express the galactic red/blue shifts measured by a distant observer, that can in principle be compared to observations provided by astronomers, in terms of three arbitrary metric functions. We also clarify the contribution of the rotation and the dragging of inertial frames due to the gravitational field. We further point out the difficulties we have when determining the metric functions from observations without making reference to any theory of gravitational interactions, and comment on some possible ways to overcome them.

Journal of Physics: Conference Series **545** (2014) 012006 doi:10.1088/1742-6596/545/1/012006

2. Stationary Axisymmetric Rotation curves

We shall start by assuming that stars behave like test particles which follow time-like geodesics of a rotating axially symmetric space-time associated with galactic discs which possess such a symmetry. The most general line element for a space-time of this kind has the following form:

$$ds^2 = -e^{2\Phi}dt^2 + Q^2dr^2 + R^2\left[d\theta^2 + \sin^2\theta\,(d\varphi - W dt)^2\right], \tag{1}$$

where Φ, Q, R and W are all functions of r and θ. We shall also consider two observers $\mathcal{O}_e$ and $\mathcal{O}_d$ with 4-velocities u_e^μ, u_d^μ, respectively. Observer $\mathcal{O}_e$ corresponds to the light emitter (i.e., to the stars placed at a point P_e of space-time), and $\mathcal{O}_d$ represents the detector at point P_d, which is located far away from the light emitter and is ideally located at $r \longrightarrow \infty$.

We further assume that stars move on the galactic plane and, thus, the polar angle can be fixed $\theta = \pi/2$, so that $u_e^\mu = (U^t, U^r, 0, U^\varphi)_e$, where $U^\mu = \dot{x}^\mu$ and the dot stands for derivation with respect to the proper time of the particle (of the star).

On the other hand, the 4-velocity of the detector located "far away" from the source, in a stationary axisymmetric background, is given by $u_d^\mu = (U^t, U^r, 0, U^\varphi)_d$ in the language of the above mentioned coordinates and conventions. In this expression the U^φ component of the 4-velocity u_d^μ accounts for the dragging of the observer at P_d due to the rotation of the galaxy. This effect must be taken into account when considering the measurement of red/blue shifts of light signals emitted in our own galaxy or in galaxies "close" to ours, for instance. However, if the studied red/blue shifts correspond to galaxies located "far away" from us, so that we can neglect the dragging effect, then we can consider that the detector is static, i.e., that the $\mathcal{O}_d$'s 4-velocity is tangent to the static Killing field $\frac{\partial}{\partial t}$, and hence its 4-velocity is given by $u_d^\mu = (U^t, U^r, 0, 0)_d$. Later on we shall give a qualitative estimation of whether we can consider this ideal limit (when the dragging effect can be neglected) in terms of the contribution of the angular velocity of a given galaxy to the measured red/blue shift.

We further normalize the 4-velocity as usual ($u^\mu u_\mu = -1$), a condition which renders the following relation

$$-1 = g_{tt}\left(U^t\right)^2 + g_{rr}\left(U^r\right)^2 + g_{\varphi\varphi}\left(U^\varphi\right)^2 + 2g_{t\varphi}U^tU^\varphi \tag{2}$$

The stationary axisymmetric metric (1) possesses two commuting Killing vectors: the time-like $\varepsilon^\mu = (1, 0, 0, 0)$ and the rotational one $\psi^\mu = (0, 0, 0, 1)$. The corresponding conserved quantities are the energy and the angular momentum per unit of mass at rest of the test particle and read

$$E = -g_{\mu\nu}\varepsilon^\mu u^\nu = -\left(g_{tt}U^t + g_{t\varphi}U^\varphi\right), \tag{3}$$

$$L = g_{\mu\nu}\psi^\mu u^\nu = g_{\varphi\varphi}U^\varphi + g_{t\varphi}U^t. \tag{4}$$

These relations are useful to obtain the expressions for U^t and U^φ in terms of the metric and these conserved quantities:

$$U^t = \frac{Eg_{\varphi\varphi} + Lg_{t\varphi}}{g_{t\varphi}^2 - g_{tt}g_{\varphi\varphi}} = \frac{E - LW}{e^{2\Phi}}, \tag{5}$$

$$U^\varphi = -\frac{Eg_{t\varphi} + Lg_{tt}}{g_{t\varphi}^2 - g_{tt}g_{\varphi\varphi}} = \frac{L}{R^2} + \frac{(E - LW)\,W}{e^{2\Phi}}. \tag{6}$$

By introducing these 4-velocity components (5) and (6) in the line element (2) we get

$$g_{rr}\left(U^r\right)^2 = EU^t - LU^\varphi - 1 \tag{7}$$

or, after multiplying by $e^{2\Phi}$ we equivalently have

$$e^{2\Phi}Q^2\left(U^r\right)^2 + e^{2\Phi} + \frac{L^2e^{2\Phi}}{R^2} - (E - LW)^2 = e^{2\Phi}Q^2\left(U^r\right)^2 + V_{eff}(E, L, g_{\mu\nu}) = 0 \tag{8}$$

This equation resembles an energy conservation law for a non-relativistic particle with position dependent mass moving in an effective potential that depends on E and L. This equation can be further reduced by considering that the orbits of stars are circular $U^r = 0$:

$$V_{eff} = e^{2\Phi} + \frac{L^2 e^{2\Phi}}{R^2} - (E - LW)^2 = 0 \tag{9}$$

Hence, once we have deleted the kinetic energy component in (8) one gets the expression for the effective potential on which the star moves (9). Thus, by deriving it with respect to the radial coordinate and setting the result to zero, we basically impose the minimum condition on the effective potential, a necessary condition to have circular orbits for the stars. This condition leads to the following expression for E:

$$E = LW - \frac{\left(e^{2\Phi} + \frac{L^2 e^{2\Phi}}{R^2}\right)'}{2LW'} \tag{10}$$

whereas from equation (9) we get the following relation:

$$E = LW + e^{\Phi}\sqrt{1 + \frac{L^2}{R^2}} \tag{11}$$

From both of these equations we see that the second term in the rhs of (10) must be positive definite, indicating that the numerator and the denominator must have opposite sign and leading to the following restriction:

$$\left(e^{2\Phi} + \frac{L^2 e^{2\Phi}}{R^2}\right)' \gtrless 0 \qquad \text{and} \qquad W' \lessgtr 0, \tag{12}$$

where the last two inequalities must be satisfied simultaneously in the sense that the numerator and the denominator of the aforementioned term must have opposite sign for positive L.

Moreover, from (10) and (11) we obtain the expression for the angular momentum L:

$$L_{\pm} = \sqrt{\frac{(W')^2 \pm |W'|\sqrt{(W')^2 + \frac{2R'(e^{2\Phi})'}{R^3}} - \frac{(e^{2\Phi})'\left(\ln\frac{e^{\Phi}}{R}\right)'}{R^2}}{\frac{2}{R^2}\left\{\left[\left(\frac{e^{\Phi}}{R}\right)'\right]^2 - (W')^2\right\}}}, \tag{13}$$

where the following restrictions must be fulfilled in order to have a real angular momentum:

$$(i) \qquad (W')^2 \geq -\frac{2R'\left(e^{2\Phi}\right)'}{R^3}, \qquad \text{equivalently,} \qquad (W')^2 \geq \left(e^{2\Phi}\right)'\left[(R)^{-2}\right]', \tag{14}$$

this condition makes real the square root of the second term in the radicand's numerator; and

$$(ii) \qquad (W')^2 \pm |W'|\sqrt{(W')^2 + \frac{2R'\left(e^{2\Phi}\right)'}{R^3}} - \frac{\left(e^{2\Phi}\right)'\left(\ln\frac{e^{\Phi}}{R}\right)'}{R^2} \gtrless 0, \tag{15}$$

$$\text{or} \qquad \left|\left(\frac{e^{\Phi}}{R}\right)'\right| \gtrless |W'|, \tag{16}$$

Journal of Physics: Conference Series **545** (2014) 012006 doi:10.1088/1742-6596/545/1/012006

where now the last two inequalities must have the same sign simultaneously, either positive or negative.

On the other side, for the energy E we have from (11):

$$E_{\pm}=e^{\Phi}\sqrt{1+\frac{(W')^2\pm|W'|\sqrt{(W')^2+\frac{2R'(e^{2\Phi})'}{R^3}-\frac{(e^{2\Phi})'\left(\ln\frac{e^{\Phi}}{R}\right)'}{R^2}}}{2\left\{\left[\left(\frac{e^{\Phi}}{R}\right)'\right]^2-(W')^2\right\}}}$$

$$+W\sqrt{\frac{(W')^2\pm|W'|\sqrt{(W')^2+\frac{2R'(e^{2\Phi})}{R^3}-\frac{(e^{2\Phi})'\left(\ln\frac{e^{\Phi}}{R}\right)'}{R^2}}}{\frac{2}{R^2}\left\{\left[\left(\frac{e^{\Phi}}{R}\right)'\right]^2-(W')^2\right\}}}. \tag{17}$$

Moreover, in order for the circular orbits of stars to be stable, we also need the second derivative of the effective potential to be positive:

$$\left[e^{2\Phi}\left(1+\frac{L^2}{R^2}\right)\right]''+2L\left(E-LW\right)W''-2L^2\left(W'\right)^2>0, \tag{18}$$

where L and E must be replace by their expressions (13) and (17), respectively. It is worth noticing that relations (13) and (17), as well as the condition (18), can be reduced to the results obtained for the static spherically symmetric approach reported in [1] and [2] when $\theta=\pi/2$, $W=0$ and $R=r$ under a simple coordinate transformation. For instance, by taking $N=e^{\Phi}$ we have:

$$E^2=\frac{N^2}{1-r\partial_r N/N}, \quad L^2=\frac{r^3\partial_r N/N}{1-r\partial_r N/N} \tag{19}$$

and

$$E=\frac{e^{\Phi}}{\sqrt{1-r\Phi'}}, \quad L=\frac{r\sqrt{r\Phi'}}{\sqrt{1-r\Phi'}}, \tag{20}$$

which are precisely the expressions obtained by [1] and [2], respectively, for the energy and angular momentum.

It is well known that rotation curves of spiral galaxies are inferred from the red and blue shifts of the radiation emitted by stars that move in (nearly) circular orbits around the central region of the galaxy [4]–[7] and that light signals travel on null geodesics with tangent 4-momentum k^{μ}. Here we shall make the assumption that k^{μ} are restricted to lie in the galactic plane $\theta=\pi/2$, and we shall evaluate the frequency shift of a light signal emitted from a star in a circular orbit represented by $\mathcal{O}_e$ and detected by an observer represented by $\mathcal{O}_d$. Moreover, we shall suppose that the galactic disc is edge-on directed towards the observer, a fact which implies that $k^{\theta}=0$.[1]

The frequency shift associated to the emission and detection of light signals is given by

$$1+z=\frac{\omega_e}{\omega_d}, \tag{21}$$

where the frequency of a photon measured by an observer with proper velocity $u^{\mu}_C|_{P_C}$ reads

$$\omega_C=-k_{\mu}u^{\mu}_C|_{P_C}, \tag{22}$$

[1] This fact is also taken into account by astronomers when reporting the corresponding total blue or red shifts since they substract the contribution coming from the inclination of the galactic disc from the plane $\theta=\pi/2$.

Journal of Physics: Conference Series **545** (2014) 012006 doi:10.1088/1742-6596/545/1/012006

and the index C refers to the emission (e) or detection (d) at the corresponding space-time point P_C. When the observer is comoving with the particle we have

$$\omega_e = -k_\mu u_e^\mu \ . \tag{23}$$

Thus, a photon emitted at point P_C in the galactic plane possesses a 4-momentum $k_C^\mu = \left(k^t, k^r, 0, k^\varphi\right)_C$. The corresponding conserved quantities along the null geodesics of light signals read

$$E_\gamma = -g_{\mu\nu}\varepsilon^\mu k^\nu = -\left(g_{tt}k^t + g_{t\varphi}k^\varphi\right) , \tag{24}$$

$$L_\gamma = g_{\mu\nu}\psi^\mu k^\nu = g_{\varphi\varphi}k^\varphi + g_{t\varphi}k^t , \tag{25}$$

whereas the normalization condition for the 4-momentum $k^\mu k_\mu = 0$ is

$$0 = g_{tt}\left(k^t\right)^2 + g_{rr}\left(k^r\right)^2 + g_{\varphi\varphi}\left(k^\varphi\right)^2 + 2g_{t\varphi}k^t k^\varphi \tag{26}$$

and leads to the following expression for the k^r in terms of the metric, the conserved quantities L_γ, E_γ, and k^t and k^φ:

$$g_{rr}\left(k^r\right)^2 = E_\gamma k^t - L_\gamma k^\varphi = \frac{g_{\varphi\varphi}E_\gamma^2 + 2g_{t\varphi}E_\gamma L_\gamma + g_{tt}L_\gamma^2}{g_{t\varphi}^2 - g_{tt}g_{\varphi\varphi}}. \tag{27}$$

There are two frequency shifts which correspond to the maximum and minimum values of ω_e associated with light propagation in the same and the opposite direction of the motion of the signal emitter, in other words, the frequency shifts of a receding and an approaching star, respectively. As we shall see later, these max/min values of the frequency shifts are reached for stars whose position vector $\mathbf{r}$, with respect to the galactic center, is perpendicular to the detector's line of sight, i.e., along the plane where $k^r = 0$ for the observer [2].

From the constant character along the geodesics of the product of the Killing vector field ε^μ with a geodesic tangent (24), together with the frequency definition (22) and taking into account that at ($r \longrightarrow \infty$), the 4-velocity of the detector is given by $u_d^\mu = (U^t, 0, 0, U^\varphi)_d$, we get the following expressions for the detector's frequency

$$\begin{aligned} \omega_d &= -k_\mu u_d^\mu = -\left(g_{tt}U^t k^t + g_{t\varphi}U^t k^\varphi + g_{\varphi t}U^\varphi k^t + g_{\varphi\varphi}U^\varphi k^\varphi\right)\big|_d \\ &= \left(E_\gamma U^t - L_\gamma U^\varphi\right)\big|_d \end{aligned} \tag{28}$$

and the emitter's frequency

$$\begin{aligned} \omega_e &= -k_\mu u^\mu\big|_e = -\left(g_{tt}U^t k^t + g_{t\varphi}U^t k^\varphi + g_{\varphi t}U^\varphi k^t + g_{\varphi\varphi}U^\varphi k^\varphi + g_{rr}U^r k^r\right)\big|_e \\ &= \left(E_\gamma U^t - L_\gamma U^\varphi - g_{rr}U^r k^r\right)\big|_e \end{aligned} \tag{29}$$

Therefore, with the aid of the frequency shift (21) we have for arbitrary star orbits

$$1 + z = \frac{\omega_e}{\omega_d} = \frac{\left(E_\gamma U^t - L_\gamma U^\varphi - g_{rr}U^r k^r\right)\big|_e}{\left(E_\gamma U^t - L_\gamma U^\varphi\right)\big|_d} \ . \tag{30}$$

Since the star orbits we are considering are circular, then

$$1 + z = \frac{\omega_e}{\omega_d} = \frac{\left(E_\gamma U^t - L_\gamma U^\varphi\right)\big|_e}{\left(E_\gamma U^t - L_\gamma U^\varphi\right)\big|_d} = \frac{U_e^t - bU_e^\varphi}{U_d^t - bU_d^\varphi} \ , \tag{31}$$

where we have introduced $b \equiv \frac{L_\gamma}{E_\gamma}$ – the impact parameter at infinity, and, hence, $|b|$ represents the radial distance at any side of the observed center of the galaxy, where $b = 0$. Since we shall consider red/blue shifts either side of the central value $b = 0$, it is convenient to express (31) as

$$1 + z_\epsilon = \frac{\left(U^t - \epsilon|b|U^\varphi\right)\big|_e}{\left(U^t - \epsilon|b|U^\varphi\right)\big|_d} \,, \tag{32}$$

where $\epsilon = \pm 1$, and to compute the following quantity:

$$1 + z_c = \frac{U_e^t}{U_d^t} = \frac{\left(\frac{E-LW}{e^{2\Phi}}\right)\big|_e}{\left(\frac{E-LW}{e^{2\Phi}}\right)\big|_d} \,, \tag{33}$$

where L and E are given by (13) and (17), respectively. This is the gravitational red shift of the center of the galaxy measured by an observer located at $r \longrightarrow \infty$. Since in order to consider red/blue shifts either side of the central value $b = 0$, we need to subtract this quantity from (32), we define red and blue shifts as follows

$$Z_{red} \equiv z_+ - z_c = \frac{\left(U_e^t U_d^\varphi - U_d^t U_e^\varphi\right)}{U_d^t \left(U_d^t - |b| U_d^\varphi\right)} |b| \,, \tag{34}$$

$$Z_{blue} \equiv z_c - z_- = \frac{\left(U_e^t U_d^\varphi - U_d^t U_e^\varphi\right)}{U_d^t \left(U_d^t + |b| U_d^\varphi\right)} |b| \,, \tag{35}$$

corresponding to receding and approaching stars, respectively. It is obvious that now $Z_{red} \neq Z_{blue}$ (since $z_+ - z_c \neq z_c - z_-$). These quantities are the general red and blue shifts, respectively, experimented by light signals traveling along null geodesics and emitted by circularly orbiting stars around the center of a galaxy to a distant observer. It is worth mentioning that in this formula we have dropped the gravitational red shift of the center of the galaxy, a fact that it is indeed taken into account by astronomers when reporting their observed data.

In the special case in which the detector is located "far away" enough from the source of information, i.e., if the contribution of the dragging of the detector's inertial frame due to the rotation of the system, encoded in U_d^φ, is negligible in comparison to the contribution coming from the U_d^t component, in other words, if $U_d^\varphi << U_d^t$, then the detector can be considered static at $r \longrightarrow \infty$ and its 4-velocity will be given by $u_d^\mu = e^{2\Phi(\infty)}\delta_t^\mu$. This fact can be understood as well from the following analysis: since we are considering the limit in which $U_d^\varphi << U_d^t$, and taking into account that $u_d^\mu = \frac{dx^\mu}{ds}\big|_d$ then

$$\frac{U_d^\varphi}{U_d^t} = \frac{d\varphi}{dt} \equiv \Omega_d << 1, \tag{36}$$

where s is the proper time of the orbiting particle (star) and Ω_d is the angular velocity measured by the detector at $r \longrightarrow \infty$.

Thus, in the special case in which $U_d^\varphi << U_d^t$ (or $\Omega_d << 1$,) the general red/blue shifts (34) and (35) reduce to the effective (quasi-static) red/blue shift defined by the following expression:

$$Z = -\frac{U_e^\varphi}{U_d^t}|b| = -\frac{\left[\frac{L}{R^2} + \frac{(E-LW)W}{e^{2\Phi}}\right]\Big|_e}{\left(\frac{E-LW}{e^{2\Phi}}\right)\big|_d}|b| \,, \tag{37}$$

Journal of Physics: Conference Series **545** (2014) 012006 doi:10.1088/1742-6596/545/1/012006

which is symmetric with respect to the center of the galaxy as in [1] and [2]. It is worth noticing that even when the dragging of the detector's inertial frame is neglected, the differential rotation encoded in W, still plays a nontrivial role in the relation (36). Hence we have

$$Z^2 = \frac{\left[\frac{L}{R^2} + \frac{(E-LW)W}{e^{2\Phi}}\right]^2\Big|_e}{\left(\frac{E-LW}{e^{2\Phi}}\right)^2\Big|_d} b^2 . \tag{38}$$

We further need to take into account the light bending due to the gravitational field generated by the rotating galaxy, in other words, we need to construct a mapping between the impact parameter b and the location of the star r given by its vector position $\mathbf{r}$ with respect to the center of the galaxy, i.e., the mapping $b(r)$. Following [1] and [2], we shall choose the maximum value of Z^2 at a fixed distance from the observed center of the galaxy (at a fixed b). From (38) it follows that if the function factor that multiplies b^2 is monotone decreasing with increasing r, then the maximum observed value of Z^2 corresponds to the minimum value of r along the null geodesic of the photons. This minimum value of r corresponds to the position of the star either side of the center of the galaxy, lying on the plane perpendicular to the detector's line of sight, i.e., on the plane where $k^r = 0$ for the observer located at $r \longrightarrow \infty$. Thus, from (38) it follows that the squared impact parameter b^2 must also be maximized; this quantity can be calculated from the geodesic equation of the photons (or, equivalently, from the $k^\mu k_\mu = 0$ relation taking into account that $k^r = 0$) and is given by

$$b_\pm = \frac{-g_{t\varphi} \pm \sqrt{g_{t\varphi}^2 - g_{tt}g_{\varphi\varphi}}}{(g_{tt})} = -\frac{R^2W \pm Re^\Phi}{e^{2\Phi}}, \qquad \text{hence} \qquad b_\pm^2 = \frac{\left(R^2W \pm Re^\Phi\right)^2}{e^{4\Phi}}. \tag{39}$$

Since we look for the maximum value of b, then we should choose either b_+ or b_-, depending on the sign and magnitude of the product RW with respect to e^Φ, in such a way that b is maximum. Finally, the monotone decreasing condition of the factor that multiplies b^2 in (38) imposes the following restriction

$$\left[\ln\left(\frac{L}{R^2} + \frac{(E-LW)W}{e^{2\Phi}}\right)_e\right]' < \left[\ln\left(\frac{E-LW}{e^{2\Phi}}\right)_d\right]' . \tag{40}$$

Thus, the mapping $b(r)$, responsible for the gravitational light bending, is given by (39) under the condition (40).

We should also mention here that the relations (39) and (40) reduce to the results obtained in [1] and [2] when $W = 0$ and $R = r$:

$$b^2 = \frac{r^2}{e^{2\Phi}} \qquad \text{and} \qquad \left(\ln\frac{L}{r^2}\right)' < \left(\ln\frac{E}{e^{2\Phi}}\right)' , \tag{41}$$

respectively.

As mentioned above, observations are reported as

$$Z = v(b) - v(b=0) \tag{42}$$

where Z is given by (38), supplemented by the light bending mapping (39).

3. Discussion

If one succeeds in determining the three unknown metric functions Φ, R and W from observational data, this would imply that the dynamics of light signals is determined by the geodesics of a stationary axisymmetric metric, independently of the assumed dynamics of the geometry (of a concrete theory of gravitational interactions) or of the nature of dark matter, in the case that the latter is needed. However, the task of solving for Φ, R and W is not trivial at all compared to the spherically symmetric case considered in [1] and [2] where there was just one unknown metric function to be determined.

A way of determining the above mentioned three arbitrary metric functions from observations consists in making use of the empiric Persic and Salucci's universal formula for rotation curves in the halo region of a galaxy [8] (see also [30]):

$$Z_i = \frac{\alpha_i b^2}{b^2 + \beta_i}, \tag{43}$$

where the constants α_i and β_i correspond to the description of the central, red and blue shifts defined in (33)–(35) and labeled by the index i. These constants are determined from observations of a given galaxy.

An alternative way to proceed is to consider just the red and blue shifts, (34) and (35), related to this formula and make use of the relation (10) (or (11) if possible) in order to complete the system of equations for Φ, R and W. Thus, in principle, we can conform a system of three nonlinear differential equations of first order for three unknown metric functions. From the expressions found for the red/blue shifts it is obvious that the difficulties that arise in solving such a system are not trivial at all.

There is another kind of universal rotation curve formula for flat spiral galaxies that can be adjusted to observed data [31] and can be used to determine Φ, R and W upon setting it to the red/blue shift language. Moreover, both of these universal rotation curve formulas could be used in a combined way in order to determine the metric functions Φ, R and W from observations for one and the same sample of galaxies.

One more way to find these functions consists of approaching the galactic rotation curves problem in the post-Newtonian approximation in the spirit of [32], where the authors constructed analytical models that allow one to compute the first general relativistic corrections to the matter distributions and gravitational potentials for stationary systems with axial symmetry. It is worth mentioning that the main modifications in their approach appear far from the galaxy cores, a result that seems to be consistent with the predictions of [18]–[22].

We would like to point out as well that one can make a combined use of the red/blue shifts of galactic rotation curves and gravitational lensing as proposed in [33]–[34]. However, as stated in [34], till now the combined measured data come from different distance scales (red shifts): most high quality rotation curves are available for galaxies with a low to intermediate red shift of up to $z \sim 0.4$, while gravitational lenses are easier to detect at intermediate to high red shifts $z \gtrsim 0.4$. Therefore, both kinds of data are available for the same galaxy, but at different radii and, hence, they are not comparable. Thus, it is still difficult to take advantage from both sets of observations simultaneously. This situation will improve in the future when observations with a higher resolution will be carried out. In summary, more precise observational data are needed in order to benefit from the stationary axisymmetric approach to the galactic rotation curves problem within this context.

We should also mention that the presented stationary axisymmetric formalism can be applied to a wider range of astrophysical phenomena that leave the galactic framework like binary systems, accretion discs of rotating black holes and active galactic nuclei where the magnitude of the effects would be less restrictive compared to the galactic rotation curves. Some of these issues are currently being pushed forward by the authors of this contribution and will be presented soon elsewhere.

Acknowledgments

The authors thank Pedro Colín and Daniel Sudarski for fruitful and illuminating discussions and SNI for partial financial support. AHA is also grateful to the Physics Department of UAM-I for hospitality and acknowledges support from the PAPIIT-UNAM grant No. IN103413-3 *Teorías de Kaluza-Klein, inflación y perturbaciones gravitacionales.* UN acknowledges support from the CIC–UMSNH Project No. 4.8 and PROMEP.

References

[1] Nucamendi U, Salgado M and Sudarsky D 2001 *Phys. Rev.* D **63** 125016
[2] Lake K 2004 *Phys. Rev. Lett.* **92** 051101
[3] Herrera–Aguilar A, Nucamendi U, Santos E, Corradini O and Álvarez C 2013 *Mont. Not. Roy. Astron. Soc.* **432** 301
[4] Rubin V C 1983 *Science* **220** 1339
[5] Rubin V C, Ford Jr W K and Thonnard N 1980 *Astrophys. J.* **238** 471
[6] Rubin V C, Ford Jr W K, Thonnard N and Burstein D 1982 *Astrophys. J.* **261** 439
[7] Rubin V C, Ford Jr W K, Thonnard N and Burstein D 1985 *Astrophys. J.* **289** 81
[8] Persic M, Salucci P and Stel F 1996 *Mon. Not. Roy. Astron. Soc.* **281** 27
[9] Sofue Y V and Rubin V C 2001 *Ann. Rev. Astron. Astrophys.* **39** 137
[10] Komatsu E et al. 2011*Astrophys. J. Suppl.* **192** 18
[11] Salucci P, Lapi A, Tonini C, Gentile G, Yegorova I and Klein U 2007 *Mon. Not. Roy. Astron. Soc.* **378** 41
[12] Milgrom M 1983 *Astrophys J.* **270** 365
[13] Milgrom M and Sanders R H 2008 *Astrophys J.* **678** 131
[14] Boehmer C G, Harko T and Lobo F S N 2008 *Astropart. Phys.* **29** 386
[15] Mendoza S, Hernández X, Hidalgo J C and Bernal T 2011 *Mon. Not. Roy. Astron. Soc.* **411** 226
[16] Hernández X, Jiménez M A and Allen C 2012 *Eur. Phys. J.* C **72** 1884
[17] Matos T, Núñez D, Guzmán F S and Ramírez E 2002 *Gen. Rel. Grav.* **34** 283
[18] Cooperstock F I and Tieu S 2005 astro-ph/0507619
[19] Cooperstock F I and Tieu S 2005 astro-ph/0512048
[20] Cooperstock F I and Tieu S 2006 *Mod. Phys. Lett.* A **21** 2133
[21] Cooperstock F I and Tieu S 2007 *Int. J. Mod. Phys.* A **22** 2293
[22] Cooperstock F I and Tieu S 2008 *Mod. Phys. Lett.* A **23** 1745
[23] Carrick J D and Cooperstock F I 2012 *Astrophys. Space Sci.* **337** 321
[24] Korzynski M 2005 astro-ph/0508377
[25] Menzies D and Mathews G J 2006 gr-qc/0604092
[26] Zingg T, Aste A and Trautmann D 2007 *Adv. Stud. Theor. Phys.* **1** 409
[27] Cross D J 2006 astro-ph/0601191
[28] Rakic A and Schwarz D J 2008 *PoS* **IDM2008** 096
[29] Binney J and Tremaine S 2008 *Galactic Dynamics* (Princeton: Princeton University Press)
[30] Salucci P 2001 *Mon. Not. Roy. Astron. Soc.* **320** L1
[31] González G A, Plata–Plata S M and Ramos–Caro J 2010 *Mont. Not. Roy. Astron. Soc.* **404** 468
[32] Ramos–Caro J, Agón C A and Pedraza J F 2012 *Phys. Rev.* D **86** 043008
[33] Bharadwaj S and Kar S 2003 *Phys. Rev.* D **68** 023516
[34] Faber T and Visser M 2006 *Mont. Not. Roy. Astron. Soc.* **372** 136

Journal of Physics: Conference Series **545** (2014) 012007 doi:10.1088/1742-6596/545/1/012007

Stellar Stability in the Braneworld

Francisco Linares[a], **Miguel A. García-Aspeitia**[b], **L. Arturo Ureña-López**[c]

Departamento de Física, DCI, Campus León, Universidad de Guanajuato, 37150, León, Guanajuato, México

E-mail: [a]`fran2012@fisica.ugto.mx`, [b]`aspeitia@fisica.ugto.mx`, [c]`lurena@fisica.ugto.mx`

Abstract. We study the properties of astrophysical systems in a *Brane World* scenario. In particular, the consequences of local terms that emerge from extra-dimensional gravity are explored through the modified Tolman-Oppenheimer-Volkoff equation. We compute the induced changes into mass, density and compactness of objects like white dwarfs and neutron stars. The main corrections in comparison with the known results of General Relativity are given in terms of the ratio ρ_0/λ, where ρ_0 is the central energy density of the stellar distribution and λ is the brane tension. As a result of these modifications, We obtain that the compactness of both, neutron stars and white dwarf decreases with respect to the value of General Relativity.

1. Introduction.

It is well known that stellar systems are excellent laboratories to study *General Relativity* (GR), being this theory confirmed over and over again since its almost one hundred years (see for example [1]). However there are astrophysical systems that possess high energies due to their extremely high densities and pressures, so GR corrections are expected.[1]

For this reason, it is important to have modifications to GR. An extension to the Einstein theory of gravity is offered by the *Brane World* models (BW), which states that our Universe has a space-time of five dimensions. These models assume that interactions of the standard model are confined to the (3+1)-dimensional hypersurface (*brane*), while gravity can leak to the extra dimension; this is the reason why gravity is perceived weaker compared with the other fundamental interactions, as proposed by Randall-Sundrum (RS) in the models RSI and RSII [3, 4]. The effects of extra dimension can be observed in the Einstein equations, which are modified by the appearance of new terms: the local or high energy term, which depend on quadratic terms of the energy-momentum tensor, the nonlocal or geometric term provided by the Weyl tensor and finally the possible matter content in the extra dimension. The presence of these new terms, provide us with new dynamic in the study of astrophysical systems.

Based in these ideas, this work is dedicated to extend the results provided by several authors [5, 6] to the case of a polytropic equation of state (EoS). For simplicity, we exclude the nonlocal terms, preserving the quadratic part of the energy-momentum tensor. We will be interested in three regimes: high, comparable and low energy regime, whose main distinction is given by the ratio ρ_0/λ.

We will proceed as follows: In Section 2 a RSII-like scenario is established. The effective Einstein equations on the brane will arise following the methodology of [7, 8] and then in

[1] A good system to test GR was discovered few month ago as reported at [2].

Section 3, we will obtain the modified Tolman-Oppenheimer-Volkoff (TOV) equation by the local term of the extra-dimensional gravity. By using this TOV equation will be possible to analize a particular astrophysical system to study the consequences on the stellar object due to extra dimension, as shown in Section 4 where two astrophysical distributions are studied: 1.- a neutron star with constant density, which analytical expressions for pressure, the bound for the value of brane tension and the limit for gravitational collapse are obtained, and 2.- a white dwarf, which matter content is well described by a polytropic EoS, this is, a relationship between pressure P and density ρ described by $P \propto \rho^{\gamma}$, where γ is a fractional exponent. Finally, in Section 5 we discuss the results and perspectives of this work.

2. Einstein Field Equations on the brane (a mathematical review).

Let us start by writing the equations describing a (3+1) brane embedded in a five dimensional space-time using the Randall-Sundrum II (RSII) model [4, 7, 8]. We first assume that the gravitational equation of the 5D Universe has the same mathematical structure of the Einstein field equations in four dimensions,

$$G_{AB} + \Lambda_{(5)} g_{AB} = \kappa^2_{(5)}\ {}^{(5)}T_{AB}\,, \tag{1}$$

where G_{AB} denotes the Einstein tensor, ${}^{(5)}T_{AB}$ refers to the energy-momentum tensor, $\Lambda_{(5)}$ represents the cosmological constant, and $\kappa^2_{(5)}$ is the gravitational coupling, all of them in 5D. In order to write the gravitational equations of motion in the 4D brane, we use the Gauss and Codacci equations respectively,

$$ {}^{(4)}R^{\alpha}_{\beta\gamma\delta} = {}^{(5)}R^{\mu}_{\nu\rho\sigma} q^{\alpha}_{\mu} q^{\nu}_{\beta} q^{\rho}_{\gamma} q^{\sigma}_{\delta} + K^{\alpha}_{\gamma} K_{\beta\delta} - K^{\alpha}_{\delta} K_{\delta\gamma}\,, \tag{2}$$

$$ D_{\nu} K^{\nu}_{\mu} - D_{\mu} K = {}^{(5)}R_{\rho\sigma} n^{\sigma} q^{\rho}_{\mu}\,, \tag{3}$$

where the extrinsic curvature over the 4D manifold $\mathcal{M}$ is given by $K_{\mu\nu} = q^{\alpha}_{\mu} q^{\beta}_{\nu} \nabla_{\alpha} n_{\beta}$, $K = K^{\mu}_{\mu}$, and D_{μ} is the covariant differentiation of $q_{\mu\nu}$. It is important to remark that in the BW scenario, our 4D world is described by a 3-brane ($\mathcal{M}$,$q_{\mu\nu}$) within a 5D spacetime ($\mathcal{V}$,$g_{\mu\nu}$). We denote the unit vector normal to $\mathcal{M}$ by n^{α}, and the induced metric onto it by $q_{\mu\nu} = g_{\mu\nu} - n_{\mu} n_{\nu}$ [8]. After an appropriate computation, it is possible to demonstrate that the modified 4D Einstein's equation can be written as [7, 8]

$$G_{\mu\nu} + E_{\mu\nu} + \Lambda_{(4)} g_{\mu\nu} = \kappa^2_{(4)} T_{\mu\nu} + 6\frac{\kappa^2_{(4)}}{\lambda}\Pi_{\mu\nu} + 4\frac{\kappa^2_{(4)}}{\lambda} F_{\mu\nu}\,, \tag{4}$$

where

$$\Lambda_{(4)} = \frac{1}{2}\Lambda_{(5)} + \frac{\kappa^4_{(5)}}{12}\lambda^2\,, \tag{5}$$

$$\kappa^2_{(4)} = 8\pi G_N = \frac{\kappa^4_{(5)}}{6}\lambda\,, \tag{6}$$

$$\Pi_{\mu\nu} = -\frac{1}{4}T_{\mu\alpha}T^{\alpha}_{\nu} + \frac{T T_{\mu\nu}}{12} + \frac{g_{\mu\nu}}{24}(3T_{\alpha\beta}T^{\alpha\beta} - T^2)\,, \tag{7}$$

$$F_{\mu\nu} = \frac{2\ {}^{(5)}T_{AB} g^A_{\mu} g^B_{\nu}}{3} + \frac{2 g_{\mu\nu}}{3}\left[{}^{(5)}T_{AB} n^A n^B - \frac{{}^{(5)}T}{4}\right]\,, \tag{8}$$

$$E_{\mu\nu} \equiv {}^{(5)}C^{E}_{AFB} n_E n^F g^A_{\mu} g^B_{\nu}\,. \tag{9}$$

Journal of Physics: Conference Series **545** (2014) 012007 doi:10.1088/1742-6596/545/1/012007

Here, $\Lambda_{(4)}$ is the four dimensional cosmological constant, $\kappa_{(4)}$ is the 4D gravitational constant related to the Newton gravitational constant G_N, λ is the brane tension, and ${}^{(5)}C^{E}_{AFB}$ is the 5D Weyl tensor [7, 8]. Note that we will use the most-plus signature (diag$(-,+,+,+)$) for the line element, and natural units in which $c = \hbar = 1$.

3. The Tolman-Oppenheimer-Volkoff equation with high energy corrections.

Now, we turn our attention to solve the modified Einstein equations given by (4) under the following considerations: At astrophysical scales the cosmological constant on the brane is zero, *i.e.*, $\Lambda_{(4)} = 0$. This agrees with the fact that the observed cosmological constant is practically zero, so the 5D spacetime is Anti de Sitter, because Eq.(5) yields

$$\Lambda_{(4)} \sim 0 = \frac{1}{2}\left(\Lambda_{(5)} + \frac{1}{6}\kappa^4_{(5)}\lambda^2\right), \quad \Lambda_{(5)} = -\frac{1}{6}\kappa^4_{(5)}\lambda^2 \,. \tag{10}$$

It will be also assumed for simplicity that the nonlocal terms are negligible at astrophysical scales, and then $E_{\mu\nu} = 0$. This is a very strong assumption because it is expected that an observer on the brane may be able to detect nonlocal effects due to the 5D gravity. In fact, notice that by taking the relativistic limit $\lambda \to \infty$ in Eq. (4), the term containing the Weyl tensor remains. However, in special cases the symmetries of the system allow us to simplify the Weyl tensor. For example, $E_{\mu\nu} = 0$ for a 5D conformally flat spacetime [7].[2]

Without matter content in the extra dimension, we will have ${}^{(5)}T_{AB} = 0$, and this implies that $F_{\mu\nu} = 0$ [5, 7]. Also, we find that $\nabla^{\mu}T_{\mu\nu} = 0$, because Eqs. (1) and (3), together with the Israel conditions[10], imply that

$$\nabla^{\mu}T_{\mu\nu} = -2\;{}^{(5)}T_{AB}n^{A}g^{B}_{\ \nu} = 0\,, \tag{11}$$

which is consistent with the Bianchi identity in 4D gravity.[3]

With all the assumptions given previously, Eqs. (4) becomes

$$G_{\mu\nu} = 8\pi G_N T_{\mu\nu} + \frac{48\pi G_N}{\lambda}\Pi_{\mu\nu}\,. \tag{12}$$

The tensors $T_{\mu\nu}$ and $\Pi_{\mu\nu}$ will describe a perfect fluid, and their explicit forms are

$$T_{\mu\nu} = \rho u_{\mu}u_{\nu} + P h_{\mu\nu}\,, \quad \Pi_{\mu\nu} = \frac{1}{12}\rho\left[\rho u_{\mu}u_{\nu} + (\rho + 2P)h_{\mu\nu}\right]\,, \tag{13}$$

where $\rho(r)$ and $P(r)$ are the energy density and pressure respectively. To solve Eqs. (12), let us consider the following static and spherically symmetric line element,

$$ds^2 = -B(r)dt^2 + A(r)dr^2 + r^2[d\theta^2 + \sin^2(\theta)d\varphi^2]\,, \tag{14}$$

where $A(r)$ and $B(r)$ are the gravitational potentials. By writing in components the Eq.(12), and by eliminating the gravitational potentials $A(r)$ and $B(r)$, it is possible to obtain a first order differential equation in terms of the physical quantities of the star: the energy density $\rho(r)$ and pressure $P(r)$,

$$-r^2\frac{dP(r)}{dr} = \frac{G_N M(r)\rho(r)}{1 - 2G_N M(r)/r}\left[1 + \frac{P(r)}{\rho(r)}\right]\left[1 + \frac{4\pi P(r)r^3}{M(r)}\left(1 + \frac{\rho(r)}{2\lambda}\left[2 + \frac{\rho(r)}{P(r)}\right]\right)\right]\,, \tag{15}$$

[2] Solutions that contain $E_{\mu\nu} \neq 0$ can be seen at [5, 6]. For a case where $E_{\mu\nu} \neq 0$ but $\Pi_{\mu\nu} = 0$ see for instance [9].

[3] See [11] for cases where $\mathcal{F}_{\mu\nu} \neq 0, \Lambda \neq 0$.

Journal of Physics: Conference Series **545** (2014) 012007 doi:10.1088/1742-6596/545/1/012007

where the mass $M(r)$ is given by

$$M(r) = 4\pi \int_0^r dr' \rho(r') r'^2 \left[1 + \frac{\rho(r')}{2\lambda}\right] = M_0(r) + \frac{2\pi}{\lambda} \int_0^r dr' \rho^2(r') r'^2 . \tag{16}$$

Notice that we have also defined the physical mass as $M_0(r)$, which coincides with the standard GR definition of mass for a stellar distribution. The complete mass function $M(r)$ includes a correction term that depends on the brane tension λ.

Therefore, Eq. (15) is the TOV equation with corrections coming from 5D gravity. Clearly, if we take the relativistic limit $\rho/\lambda \to 0$ we obtain the TOV equation of GR, and then it is natural to consider corrections to the standard results in terms of that ratio. For instance, according to this criterion, the mass of the star configuration $M \to M_0$ in the limit $\rho/\lambda \to 0$, as can be seen from Eq.(16).

4. Inner solutions from modified TOV equation.

We will describe solutions of the modified TOV equation (15) for two astrophysical systems:

1.- Neutron star with a constant energy density: ρ =const.

2.- White dwarf with a polytropic equation of state: $P(r) = K\rho^\gamma(r)$.

The first case is of particular interest because, even if it is not realistic, their solutions may provide a useful insight on the nature of the brane corrections that modify the well known GR results. The second case may have a more realistic interpretation in terms of a white dwarf since they are objects that can be observed in nature, and therefore represent a "natural laboratory" to test the theoretical models.

4.1. The neutron star case. (A review of uniform density)

This case was studied in detail in [5], and we give here a brief review of the main results. Because the energy density is a constant quantity, we can find an exact solutions of Eqs.(16) and (15). The mass $M(r)$ from Eq.(16) is given by

$$M(r) = \int_0^r 4\pi r'^2 \rho \left(1 + \frac{\rho}{2\lambda}\right) dr' = \frac{4\pi r^3}{3} \rho \left(1 + \frac{\rho}{2\lambda}\right) , \tag{17}$$

and then Eq.(15) becomes

$$\frac{dP(r)}{dr} = -\frac{G_N M(r)\rho}{r^2} \left[1 + \frac{P(r)}{\rho}\right] \left[1 + \frac{4\pi P(r) r^3}{M(r)} \left(1 + \frac{\rho}{2\lambda}\left[2 + \frac{\rho}{P(r)}\right]\right)\right] \left[1 - \frac{2G_N M(r)}{r}\right]^{-1} . \tag{18}$$

We then obtain the following exact solution of Eq.(18),

$$\frac{P(r)}{\rho} = \frac{\sqrt{1 - 2G_N M/R} - \sqrt{1 - 2G_N M r^2/R^3}}{\sqrt{1 - 2G_N M r^2/R^3} - 3\left(\frac{1+\rho/\lambda}{1+2\rho/\lambda}\right)\sqrt{1 - 2G_N M/R}} , \tag{19}$$

where $M \equiv M(R)$ and R represent the total mass and radius of the star respectively. From Eq.(19) is possible to see that in the limit $\rho/\lambda \to 0$ we recover the well known solution of GR (see for instance[12, 13, 14]). Several cases of the solution obtained in Eq.(19) are shown in Fig.2 for different values of the ratio ρ/λ.

From condition which establishes that pressure is regular at the origin, we find that

$$\frac{G_N M_0}{R} < \frac{4}{9} \left[\frac{(3 + 3\rho/\lambda)^2 - (1 + 2\rho/\lambda)^2}{8(1 + \rho/2\lambda)(1 + \rho/\lambda)^2}\right] , \tag{20}$$

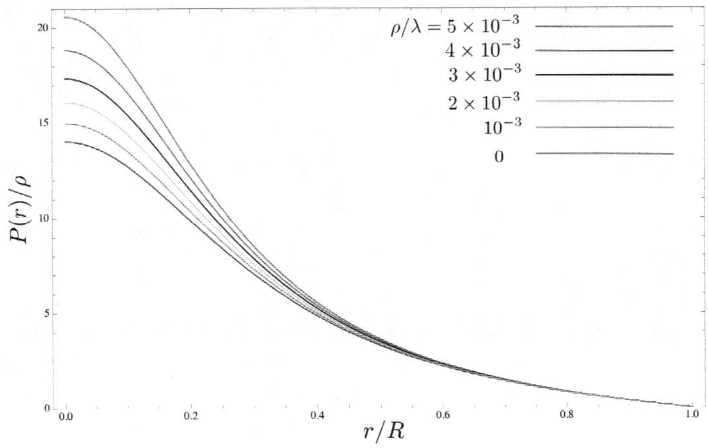

Figure 1. The radial profile of the ratio $P(r)/\rho$ for a star with a constant energy density, for different values of the ratio ρ/λ are shown. The most remarkable effect arises for the central value of the pressure, which increases for larger values of the ratio ρ/λ.

where $M_0 = 4\pi\rho R^3/3$ is the total physical mass, (see Eq.(17)). This result must be compared with the GR expression given by $G_N M/R < 4/9$, which is readily obtained in the limit $\rho/\lambda \to 0$. Following the procedure of [5], we expand Eq.(20) up to first order in ρ/λ to obtain

$$\frac{G_N M_0}{R} < \frac{4}{9} - \frac{1}{3}\left(\frac{\rho}{\lambda}\right) . \tag{21}$$

By considering the typical values of the mass, radius and mean energy density of a neutron star, it is found that the brane tension must obey the bound $\lambda > 6.320\times10^{10}\mathrm{MeV}^4$, which agrees with other reported values in [7, 5, 15, 16], and also with the bound imposed by nucleosynthesis, $\lambda \geq 1\mathrm{MeV}^4$.

4.2. The white dwarf case.

Now, we turn our attention to the case of a white dwarf. The matter content of the star will be described by a polytropic equation of state, which has the form: $P(r) = K\rho^{\gamma}(r)$, where K is a proportionality constant and γ is a fractional exponent; which value will be $\gamma = 4/3$. To solve the modified TOV equation, it is convenient to define the following dimensionless variables: $\bar{M} \equiv \sqrt{G_N}\, M(r)$, $\bar{\rho} \equiv G_N^2 \rho$, $\bar{r} \equiv r/\sqrt{G_N}$, $\bar{K} \equiv K/G_N^{2/3}$, $\bar{\lambda} \equiv G_N^2\lambda$.

Thus, the system of Eqs.(15) and (16) are written as

$$\frac{d\bar{M}(\bar{r})}{d\bar{r}} = 4\pi\bar{\rho}(\bar{r})\bar{r}^2\left[1 + \frac{\bar{\rho}(\bar{r})}{2\bar{\lambda}}\right] , \tag{22}$$

$$\frac{d\bar{\rho}(\bar{r})}{d\bar{r}} = -\frac{3\bar{M}(\bar{r})\bar{\rho}^{2/3}(\bar{r})}{4\bar{K}\bar{r}^2}\left(\frac{1+\bar{K}\bar{\rho}^{1/3}(\bar{r})}{1-2\bar{M}(\bar{r})/\bar{r}}\right)\left[1 + \frac{4\pi\bar{K}\bar{\rho}^{4/3}(\bar{r})\bar{r}^3}{\bar{M}(\bar{r})} + \frac{2\pi\bar{\rho}(\bar{r})\bar{r}^3}{\bar{M}(\bar{r})}(1+2\bar{K}\bar{\rho}^{1/3}(\bar{r}))\frac{\bar{\rho}(\bar{r})}{\bar{\lambda}}\right] . \tag{23}$$

It must be noticed that, like in the case of constant energy density discussed above, the equations of motion depend on ρ/λ, and the latter is invariant under the change of variables taken, that is, $\rho/\lambda = \bar{\rho}/\bar{\lambda}$. However, this time the energy density depends on

Journal of Physics: Conference Series **545** (2014) 012007 doi:10.1088/1742-6596/545/1/012007

the radial coordinate, and then we will define three different regimes according to the ratio $\bar{\rho}_0/\bar{\lambda}$, which compares the central value of the energy density with the brane tension. The regimes are discussed in the following subsections, where we take the value of central density as $\bar{\rho}(r=0)=\bar{\rho}_0=0.01$.

4.2.1. High energy regime $\bar{\rho}_0/\bar{\lambda} \gg 1$. The numerical solutions of Eqs.(22) and (23) for this case are shown in Fig.2. The numerical value for the brane tension is given by $\bar{\lambda}=0.001$. First of all, we notice that density profile is significantly modified, as it decreases more abruptly that in the standard GR case (without brane correction).

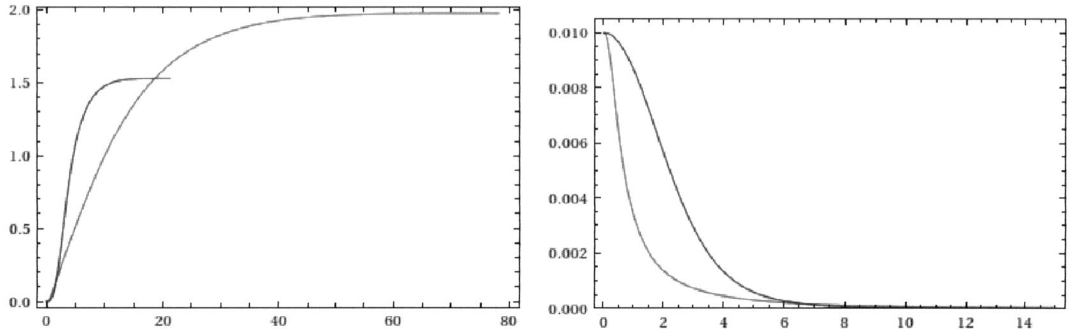

Figure 2. *Solution of the modified TOV equation in the case of a stellar configuration with a polytropic equation of state (red solid lines) and its comparison with the solution of the standard TOV equation in the pure GR case (blue solid lines). See the text for more details.*

In order to have a proper comparison between the brane case and the standard GR case, we also show the behavior of the physical mass M_0 for both cases; we only need to recall that physical mass does not contain brane corrections (see Eq.(17)). Thus, for GR we have that $M_0 = 1.53$. Is important to note that, the radio of the stellar distribution R, is given by the numerical value for which energy density and therefore, pressure vanish.

We follow the same criterion for the compactness of the star, and define the latter in terms of the physical mass M_0. The results indicate that compactness in GR is $M_0/R = 0.072$. In this way, the modifications suffered by physical mass and compactness due to extra dimensional gravity are given by

$$M_0 = \int_0^R 4\pi\bar{r}^2\bar{\rho}(\bar{r})d\bar{r} = 1.74 \ , \quad \frac{M_0}{R} = 0.022 \ . \tag{24}$$

4.2.2. Comparable energies $\bar{\rho}_0/\bar{\lambda} \sim 1$. In this case we obtain the numerical solutions shown in Fig. 3, in which case, the numerical value for the brane tension is given by $\bar{\lambda} = 0.01$. We see that the total mass of the star with effects of the extra dimension (red line) is still larger than the relativistic total mass (blue line) and the stellar radio is larger than the radio of the star without BW corrections. However these increments in both mass and radio are smaller than the high energy regime.

Also in Fig. 3, we note that density with corrections of extra dimension (red line) is less than the relativistic case (blue line). On the other hand, the physical mass M_0 and compactness in this regime are still lesser that the GR solution.

$$M_0 = 1.50 \ , \quad \frac{M_0}{R} = 0.042 \ . \tag{25}$$

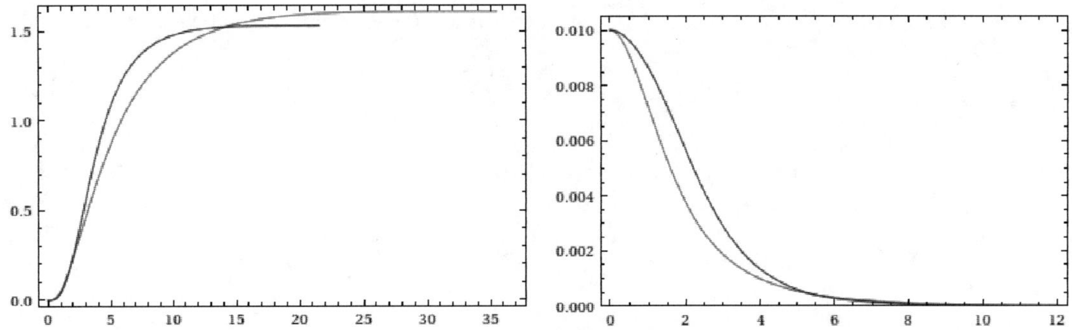

Figure 3. *The effects due to extra dimensional gravity on the mass $\bar{M}(r)$ (left) and energy density $\bar{\rho}(r)$ (right) are shown with a red line, while the blue line represents the stellar distribution without branes. In this regime is still present the effects of extra dimension, but is less than the high energy regime.*

4.2.3. Low energies $\bar{\rho}_0/\bar{\lambda} \ll 1$. Finally, we see in the following figures that is possible to obtain the relativistic solutions in this low energies regime. The numerical value for the brane tension is given by $\bar{\lambda} = 10$.

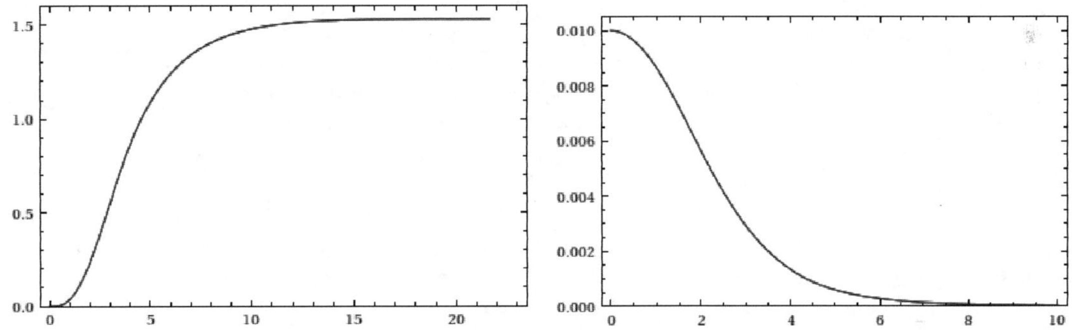

Figure 4. *The effects due to extra dimensional gravity on the mass $\bar{M}(r)$ (left) and energy density $\bar{\rho}(r)$ (right) are shown, and is clearly seen that both curves coincide. Thus, solutions of GR case are recovered.*

We can see in Fig. 4, that both curves describing the stellar mass are practically the same, *i.e.*, the total mass of the star with effects of the extra dimension (red line) has the same value that the relativistic mass (blue line) and the stellar radio is also the same. In fact, the physical mass and compactness in this regime are given by

$$M_0 = 1.53 \ , \quad \frac{M_0}{R} = 0.071 \ . \tag{26}$$

In Fig. 4, we also note that density recovers its behavior and coincides with the GR case. The following Table summarizes the results obtained above

Regime	Physical Mass M_0	Stellar Radio R	Compactness M_0/R
$\rho_0/\lambda \gg 1$	1.74	78.15	0.022
$\rho_0/\lambda \sim 1$	1.50	35.50	0.042
$\rho_0/\lambda \ll 1$	1.53	21.61	0.071

Table 1. *The numerical values obtained for the physical mass, stellar radio and compactness of the star, all of them with corrections due to the presence of extra dimension are shown. For GR case we have that physical mass, stellar radio and compactness are given by* $M_0 = 1.53$, $R = 21.45$ *and* $M_0/R = 0.072$ *respectively.*

Therefore, it is clear that is possible to recover the GR solutions, as can be seen in the low energies regime, while the effects of extra dimensions are more noticeable as the ratio $\bar{\rho}_0/\bar{\lambda}$ increases.

5. Discussion and Conclusions.

A modification to the TOV equation by considering the local term of the extra dimensional gravity was presented. This equation represents the simplest extension in these BW scenarios, because there were not considered nonlocal effects and possible matter content in the extra dimension.

Subsequently, the modified TOV equation was applied to the case of a star with constant energy density, which results are manifested in the increase of central pressure. In a more realistic case, a white dwarf was analyzed, and it was possible to see the effects that it suffers due to the extra dimension: the values for total mass and stellar radio are bigger than GR case, this effect is more noticeable in the high energies regime, while in the low energies regime these effects are practically negligible. In both neutron stars and white dwarf, the compactness decreases with respect to the GR value, but such a value is recovered when the extra dimensional term is not relevant in the equations, *i.e.*, when $\rho_0/\lambda \ll 1$.

An important fact is that all our results show that we successfully recover the relativistic solutions by taking the appropriate limit $\rho/\lambda \to 0$, which is a favorable feature of the model, if we consider it as a generalization of GR.

An extension of these results would be to consider non-local terms that arise from the projection of the Weyl tensor. This implies a new difficulty level: the system of equations given by the components of modified Einstein equation and the conservation equations is not determinated in a closed way due to the additional geometric variables. A couple of contributions that address that issue can be found in [6, 17].

6. Acknowledgements.

MAG-A acknowledges support from a CONACyT postdoctoral grant and SNI. This work was partially supported by PROMEP, DAIP, and by CONACyT México under grant 167335, the Fundación Marcos Moshinsky, and the Instituto Avanzado de Cosmología (IAC) and Beyond Standard Theory Group (BeST) collaborations.

References

[1] Will C M 2006 *Living Reviews in Relativity* **9** URL `http://www.livingreviews.org/lrr-2006-3`
[2] Ransom S M, Stairs I, Archibald A, Hessels J, Kaplan D, van Kerkwijk M, Boyles J, Deller A, Chatterjee S, Schechtman-Rook A *et al.* 2014 *Nature*
[3] Randall L and Sundrum R 1999 *Phys. Rev. Lett.* **83** 3370–3373 (*Preprint* `hep-ph/9905221`)
[4] Randall L and Sundrum R 1999 *Phys. Rev. Lett.* **83**(23) 4690–4693 URL `http://link.aps.org/doi/10.1103/PhysRevLett.83.4690`

[5] Germani C and Maartens R 2001 *Physical Review D* **64** 124010
[6] Ovalle J, Linares F, Pasqua A and Sotomayor A 2013 *Classical and Quantum Gravity* **30** 175019
[7] Maartens R and Koyama K 2010 *Living Rev. Rel* **13** 1004–3962
[8] Shiromizu T, Maeda K and Sasaki M 2000 *Phys. Rev. D* **62**(2) 024012 URL `http://link.aps.org/doi/10.1103/PhysRevD.62.024012`
[9] Rahaman F, Kalam M, DeBenedictis A, Usmani A A and Ray S 2008 *Monthly Notices of the Royal Astronomical Society* **389** 27–33
[10] Israel W 1966 *Il Nuovo Cimento B Series 10* **44** 1–14
[11] Maeda K I and Wands D 2000 *Physical Review D* **62** 124009
[12] Weinberg S 1989 *Rev. Mod. Phys* **61**
[13] Carroll S M 2001 *Living Reviews in Relativity* **4**
[14] d'Inverno R A and Harvey A 2008 *Physics Today* **46** 59–60
[15] Langlois D, Maartens R, Sasaki M and Wands D 2001 *Physical Review D* **63** 084009
[16] Garcia-Aspeitia M A 2013 *arXiv preprint arXiv:1306.1283*
[17] Ovalle J and Linares F 2013 *Phys.Rev.* **D88** 104026

Journal of Physics: Conference Series **545** (2014) 012008 doi:10.1088/1742-6596/545/1/012008

Bar formation in simulations of interacting galaxies

Juan Carlos Luna-Sánchez[1,2], **M.A. Rodríguez-Meza**[1], **A. Arrieta**[2], **and R. Gabbasov**[3]

[1]Departamento de Física, Instituto Nacional de Investigaciones Nucleares, Apdo. Postal 18-1027, México D.F. 11801, México

[3]Departamento de Física y Matemáticas, Universidad Iberoamericana, Prolongación Paseo de la Reforma 880, Lomas de Santa Fe, C.P. 01210 México, D.F.. Mexico

[3]Instituto de Ciencias Básicas e Ingenierías, Universidad Autónoma de Pachuca, Ciudad Universitaria, Carr. Pachuca-Tulancingo Km. 4.5 S/N, C.P. 42184, Pachuca, Hidalgo. Mexico

E-mail: marioalberto.rodriguez@inin.gob.mx

Abstract. In this work we present a study of interacting galaxies using N-body simulations. The initial condition of galaxies are such that they are composed of a bulge, a disc (Freeman model, with no gas), and a halo. For bulge and halo we follow the Dehnen density-pair spherical models. Galaxies are set in a parabolic encounter characterised by the impact parameter and the collision angle subtended by the planes containing each individual galactic discs. The evolution of galaxies are given in terms of the morphology (bar formation, geometry of the bar, minor and major axis length), and the kinematical bar rotation. We show how this characteristics depend on the collision geometry. The dynamics of the collision is given in terms of individual rotation curves, dispersion of velocities of the disc and mass function as functions of the distance to the center of mass of each individual galaxy.

1. Introduction

In the present work, we pursue to study bars in spiral galaxies. Observations of spiral galaxies indicate that the presence of a bar is a common feature [1, 2]. In particular, in [2] it is found that 29.4 per cent of the galaxy sample have a bar. Instabilities in isolated stellar and gaseous discs lead to bar formation; see [3] for pioneer studies and [4, 5] for a modern view. The bar formation in isolated models has been widely studied both analytically and numerically [6, 7, 8, 9, 10, 11, 12, 13]. In this paper we consider the dynamical effects of non-isolated systems which are found in clusters of galaxies. In this sense, it has been suggested that the observed bar in many spirals is the result of the gravitational interaction between two or more nearby galaxies. For instance, [14] has found that during the collision of two galaxies and between the first and the second closest approaches, the disc takes a transient bar shape. The gravitational interaction between the two galaxies gives rise to perturbations in the orbits of the stars that results in the formation of the bar.

Bar formation in stellar discs depends upon various simultaneous effects. In the case of collisions, simulations have shown that these factors are [15]: rotation curve shape, disc-halo mass ratio, perturbation force and geometry. Additionally, simulations suffer from numerical effects such as low spatial and temporal resolution, too few particles representing the system, and an approximate force model. These effects were studied in references [5, 16], where it was shown that specific parameter choices may change bar properties. Once numerical effects are controlled, we may investigate all the other model parameters, which in our case are: geometric parameters such as impact parameter and the angle between the disc planes.

In the present paper we study the formation of bars as a product of instabilities that result of the collision of two spirals. In particular we study the morphology and kinematics of bars that are formed during a binary collisions. Morphology is given by finding major and minor axis evolution. Kinematics is studied by the computation of the angular velocity of the bar. Dynamics of the disc is studied through the computations of several quantities, like rotation curve, velocity dispersion and the mass histogram as function of distance to the center of mass of each colliding galaxy.

We organise our work in the following form: In the next section we present a general way of constructing an isolated galaxy following Hernquist method [17]; and explain how to set up the parabolic collision geometry. Next, we discuss our results for various collision cases like off-axis impacts and at different angles of collision of two disc galaxies and show the resultant properties of tidally formed bars. In the final section we write our conclusions of this work.

2. Initial conditions and geometry of the collision

We use the standard procedure to construct a galaxy model with a Newtonian potential described in [5, 16]. An individual galaxy consists of a disc, halo, and bulge and its initial condition is constructed using a bulge and a Freeman disc composed of stars inmerse in a Hernquist halo model (a Dehnen's family member with $\gamma = 1$, see [18]) that acts on them gravitationally. We does not consider gas.

The spatial distribution of particles are constructed using density profiles: The bulge density profile is given by (Hernquist, 1990):

$$\rho_{\rm b}(r) = \frac{M_{\rm b} a_{\rm b}}{2\pi} \frac{1}{r(r+a_{\rm b})^3}, \tag{1}$$

and for the halo we use a Dehnen density profile with $\gamma = 0$ (Dehnen, 1993):

$$\rho_{\rm h}(r) = \frac{3M_{\rm h}}{4\pi} \frac{a_h}{(r+a_{\rm h})^4}. \tag{2}$$

We assume that disk follows the exponential profile (Freeman, 1970):

$$\rho_{\rm d}(r,z) = \frac{M_{\rm d}\alpha^2}{4\pi z_0} e^{-\alpha r} \mathrm{sech}^2\left(\frac{z}{z_0}\right). \tag{3}$$

In these equations $M_{\rm b}$, $a_{\rm b}$ and $M_{\rm h}$, $a_{\rm h}$ are the mass and length of bulge and halo respectivelly, and $M_{\rm d}$, α^{-1} y z_0 are the mass, length scale and the thickness length scale of the disk, respectivelly.

For the spherical distribution of particles, velocities are obtained using the Schwarzschild distribution,

$$f_{B,H}(v_r, v_\phi, v_\theta) \propto \exp\left[-\frac{v_r^2}{2\sigma_r^2} - \frac{v_\phi^2}{2\sigma_\phi^2} - \frac{v_\theta^2}{2\sigma_\theta^2}\right] \tag{4}$$

where σ_r, σ_ϕ, and σ_θ are the dispersion of velocities and in general they are functions of r. For an isotropic ellipsoid the above velocity distribution is the Maxwell distribution.

For a spherically simmetric mass distribution and without rotation the dispersion of velocities is obtained using Jeans' equation

$$\frac{d}{dr}\left(\rho(r)\sigma_r^2\right) + \frac{\rho(r)}{r}\left[2\sigma_r^2 - (\sigma_\theta^2 + \sigma_\phi^2)\right] = -\rho(r)\frac{d\Phi}{dr} \tag{5}$$

If the distribution of velocities is isotropic

$$\sigma_r^2 = \sigma_\theta^2 = \sigma_\phi^2 \tag{6}$$

The above equation can be integrated to give a general expression for the dispersion of velocities:

$$\sigma_r^2(r) = \frac{1}{\rho(r)} \int_r^\infty \rho(r') \frac{d\Phi}{dr'} dr' \tag{7}$$

Particles velocities can be found by inverting the equation

$$f_{b,h}(v,r) = \frac{4\pi}{(2\pi\sigma^2)^{2/3}} v^2 \exp\left[-\frac{v^2}{2\sigma_r^2}\right] \tag{8}$$

In practice it is convenient to cut the Gaussian distribution at some finite value. A natural choice is the scape velocity V_e.

For axisymmetric distribution we have that the velocity profiles for the disk are computed using the epicyclic approximation, which consists in assuming that velocity dispersions are small ($\sigma_R, \sigma_z, \sigma_\phi \ll R\omega$):

$$f_D(v_R, v_z, v_\phi) \propto \exp\left[-\frac{v_R^2}{2\sigma_R^2} - \frac{v_z^2}{2\sigma_z^2} - \frac{(v_\phi - V_0)^2}{2\sigma_\phi^2}\right] \tag{9}$$

Observations in the exterior of disk galaxies suggest that the radial dispersion is proportional to the surface radial density:

$$\sigma_R^2 \propto \exp(-\alpha\, R) \tag{10}$$

The vertical dispersion in the isothermal shell approximation is also related to the surface density of the disk:

$$\sigma_z^2 = \pi G\, z_0\, \Sigma(r) \tag{11}$$

The ratio σ_R^2/σ_z^2 is constant through the disk and is considered equal to 4, i.e.,

$$\sigma_R^2 = 4\,\sigma_z^2 \tag{12}$$

The azimuthal dispersion is simply related to radial dispersion through the epicyclic approximation for the Schwarzschild velocity distribution

$$\sigma_\phi^2 = \frac{\kappa^2}{4\omega^2} \sigma_R^2 \tag{13}$$

where ω is the angular frequency, computed from the potential

$$\omega = \frac{\partial \Phi(R)}{\partial R} \tag{14}$$

and κ is the epicyclic frequency defined by

$$\kappa^2(R) = 4\omega^2(R) + R\frac{d}{dR}\left[\omega^2(R)\right] \tag{15}$$

For an exponential surface density profile, the azimuthal drift velocity is given approximately by

$$V_0^2 = V_c^2 + \sigma_R^2 - \sigma_\phi^2 - 2\alpha R \tag{16}$$

Table 1. Parameters of the galaxy model. The system of units is such that the unit of mass, length, time, and velocity are: $2.2 \cdot 10^{11}$ $M_\odot$, 40 kpc, 0.2558 Gyr, 153 km/s, respectively.

Component	Mass	Number of particles	Cutoff radius	Scale-length
Bulge	0.0025	1024	0.308	0.008
Disc	0.1017	29491	0.5	0.045
Halo	1.6	245760	11.55	0.3

Table 2. Geometry of the numerical experiments.

ID model	Mass proportion	Impact parameter p	Collision angle	Prograde/Retrograde
IC602	1:1	0.1	30°	N/A
IC60255	"	"	0°	Prograde
IC60266	"	"	0°	Retrograde
IC6027	"	"	45°	N/A
IC6028	"	"	90°	N/A
IC6034	"	0.3	30°	N/A
IC6035	"	"	0°	Prograde
IC6036	"	"	0°	Retrograde
IC6037	"	"	45°	N/A
IC6038	"	"	90°	N/A
IC6039	"	0.6	30°	N/A
IC6040	"	"	0°	Prograde
IC6041	"	"	0°	Retrograde
IC6042	"	"	45°	N/A
IC6043	"	"	90°	N/A

where $V_c^2 = R\omega$ is the azimuthal circular velocity of the disk.

Once velocity dispersions are computed, the velocity components of particles in the disk can be found by inverting the above Gaussian distribution which includes the drift velocity V_0.

Finally, the galaxy is built, numerically, using a Monte Carlo procedure by choosing six random numbers that when transformed according to the corresponding spatial and velocity distribution functions give us the vector position and the vector velocity of each particle. This is repeated N times with N the number of particles to use in the simulation. The number of particles in each component are assigned in proportion to their masses (see table 1, where values for other parameters are given). This initial galaxy condition is relaxed up to a time of 3.0 in units of the code (0.7674 Gyrs). The geometry of the collision is such that we set both galaxies at parabolic collision orbit defined by the impact parameter p and the angle between disc planes. All experiment runs were performed with smoothing lengths $\varepsilon_b = \varepsilon_h = 0.003$, $\varepsilon_d = 0.004$ (bulge, halo, and disc), time step $\delta t = 0.01$ (but this is the maximum step size, we have using an adaptive time step scheme). In table 2 we show the experiments that we will discuss in the next section.

3. Results

All collision models have N =552,550 total particles. We consider three sets of simulations followed up to time $t = 1.56$ Gyr. The sets are defined by the impact parameter. First set is for $p = 0.1$, the second is for $p = 0.3$, and the third is for $p = 0.6$. The proportion of galaxy masses is 1:1. See table 2 for more details. In order to obtain reasonable simulations, conservation of

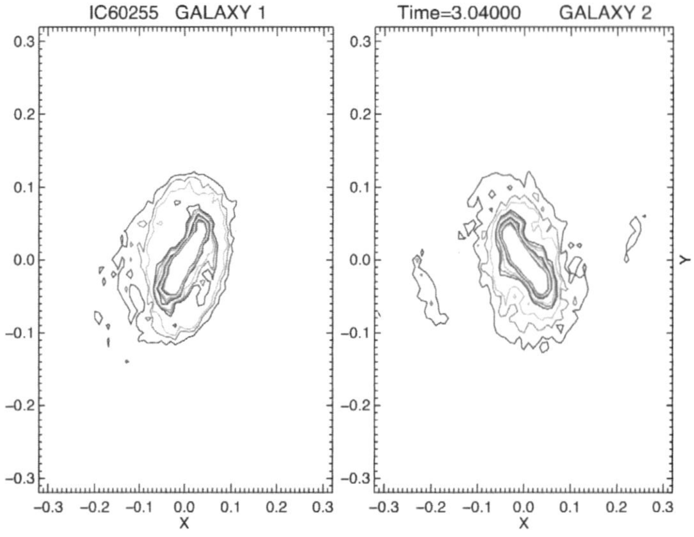

Figure 1. Contour density map of a snap shot of one of the collision model IC 60255.

total energy and total angular momentum were computed. Total energy is the sum of all kinetic and potential energies of the individual particles (halo, bulge or disc) of both galaxies. And similar for the total angular momentum. We compute at the beginning of the simulation total energy and total angular momentum values and they are our reference values, and then through all the simulation time we keep track of the values of the energy and angular momentum. At the end of the simulation we compute the evolution of the relative differences: $|(E(t) - E_0)/E_0|$ and $|(L(t) - L_0)/L_0|$. Where E_0 and L_0 are total energy and magnitud of total angular momentum at the beginning of the simulation and $E(t)$ and $L(t)$ are the corresponding quantities as functions of time. All models show a good energy and angular momentum conservation. The relative error is at most ≈ 0.4 % for the total energy and ≈ 0.8 % for the total angular momentum. We have not followed a convergence criteria in the sense of studying the influence of the numerical parameters: smoothing potential parameter, number of particles N or the time step size. This study was done in reference: [16]. Here we use numerical parameters values that are consistent with the criteria found reliable in that reference.

To detect the presence of a bar in a galaxy and characterise quantitatively its amplitude, we use the following method. We compute isodensity curves of the disc in a 128×128 mesh, i.e., a contour map of the density field around the center of mass of the disc. We tune the contour map to have a good resolutions map (see figure 1). With this contour map we scan the contour lines following several radial trajectories starting at the center of mass. The longest trajectory that intersects the first set of contour lines and that is in the same direction of the gradient at that point is the major axis. The size of this trajectory is the size of the major axis. Now, assuming that the bar is rectangular in shape we follow a trajectory perpendicular to the one that defines the major axis, and the intersection with the same set of contour lines will define the minor axis and we will obtain its size. This method allows us to detect any non-axisymmetric deformations such as bars in the disc plane. We have found traces of a bar for times between the first and second encounters. In this time interval we compute the angular velocity of the bar by computing the rate of change of the angle between the major axis and the x-axis.

In figure 2 we show the evolution of the center of separation for the three impact parameter values. The behaviour of the this curve is the same for all simulations for a given impact

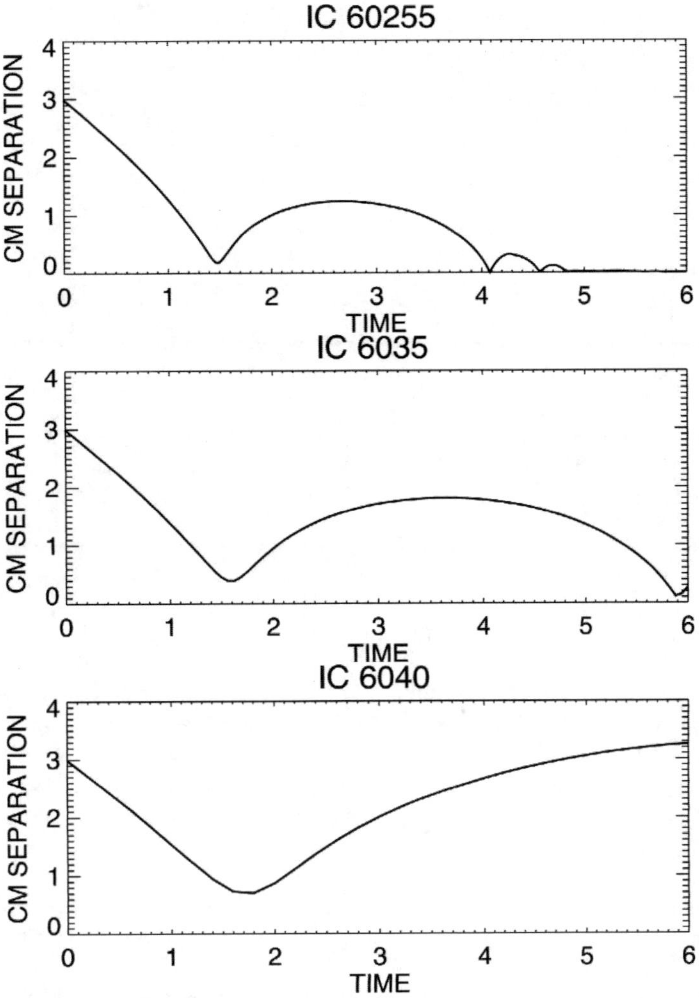

Figure 2. Evolution of the center of mass separation. Upper figure $p = 0.1$, figure in the middle $p = 0.3$, and lower figure $p = 0.6$.

parameter. This is so because the dominant component is the halo, which is spherically symmetric and the same for each galaxy. The first and second encounters, for $p = 0.1$, are, approximately, at time 0.38 Gyr and at 1.05 Gyr, respectively. Increasing impact parameter increase the times of the first and second encounters. For all the numerical experiments we look for a bar formation in each of the galaxies for times between the first and second encounter and we analyse the morphology and kinematic of these bars at five times, chosen uniformly in that time interval. The dynamics of the galaxies are studied through the computation of the rotation curves, velocity dispersion and histogram of mass of each of the colliding galaxies, at small set of times before the first encounter and at times chosen in the same way as the ones chosen for analysing the morphology and the kinematics.

In figure 3 we show the morphology and kinematics of the bar formed in the galaxy 1 with the disc in x-y plane. Morphology of the bar (size of the major and minor axes) is almost the same for $p = 0.3$ and $p = 0.6$. For $p = 0.1$ the major (minor) axis is longer (smaller) than the

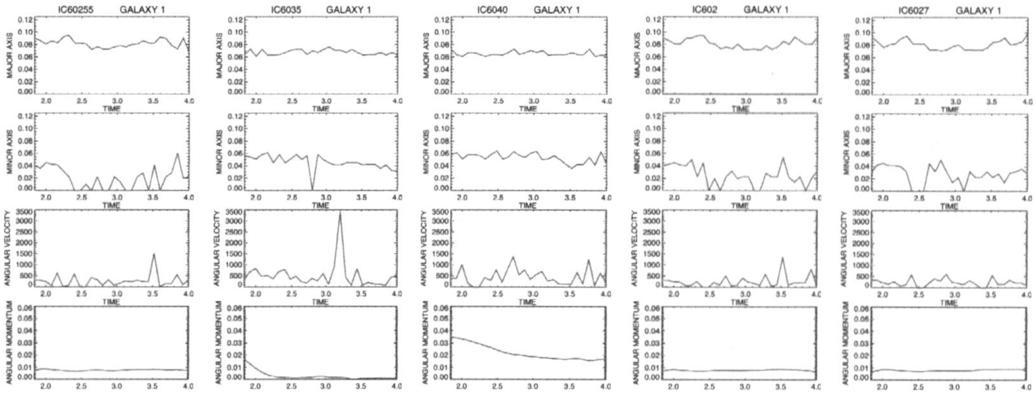

Figure 3. Morphology and kinematics of the bar formed in galaxy 1. Three models: $p = 0.1$, $p = 0.3$, and $p = 0.6$. In the first three panels both galaxies have the disc in the x-y plane. In the last two panels we have discs at different angle, first the collision angle was 30°and for the last one 45°.

ones for $p = 0.3$ and $p = 0.6$. On the average, bars for $p = 0.3$ and $p = 0.6$ move faster than the case of $p = 0.1$. Whereas changing the angle between disc planes does not change morphology or kinematics for $p = 0.1$. The angular momentum of the disc of galaxy 1 increases when impact parameter increases. In the cases in which impact parameter is $p = 0.3$ and $p = 0.6$ we have observed that changing the angle among galactic discs do not have influence on bar morphology similar as the case with $p = 0.1$. Bar's axis size is not exceeding values of 0.08, for the mayor axis, and 0.04 for the minor axis. The angular momentum, for the same cases, has values not exceeding 0.005, which is in the same range as the case, $p = 0.1$.

In figure 4 we show the dynamics of the disc in galaxy 1 (with the disc in x-y plane). The rotation curve increases for $p = 0.3$ and $p = 0.6$ with respect to the case of $p = 0.1$. The oposite behaviour occurs for the dispersion of velocities. Dynamics of the disc, for a given impact parameter, is independent of angle of collision.

We have also observed that it is clear from the results analysis that there is a correlation between morphology of the bar and dispersion of velocities of the disc. When dispersion is higher bar is longer and thinner.

4. Conclusions

Our simulations have shown that tidal forces are an efficient mechanism to generate bars in spirals. Morphology, kinematics and dynamics of the barred spirals that result from the collision do not show strong dependence on the geometry of the collision.

There is an indication of a correlation between morphology of the bar and dispersion of velocities of the disc. When the dispersion of velocity increases (in general) in the spatial range we measured, major axis increases and minor axis decreases.

References

[1] Elmegreen B G and Elmegreen D M 1983 ApJ **267** 31
[2] Masters K L, et al. 2011 Mon. Not. R. Astron. Soc. **411** 2026
[3] Toomre A 1964 ApJ **139** 1217
[4] Barnes J E 1998 In *Galaxies: Interaction and Induced Star Formation* ed D Friedly, L Martinet and D Pfenniger (Berlin: Springer-Verlag) p 275

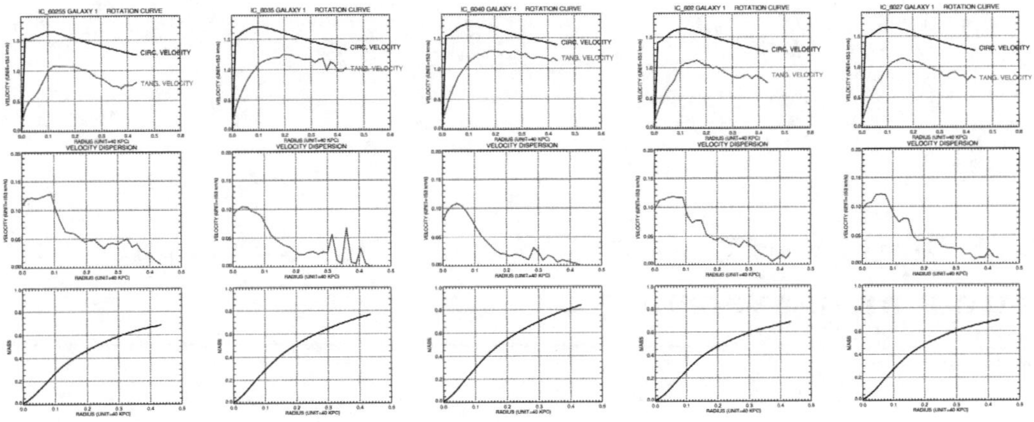

Figure 4. Dynamics of the bar formed in galaxy 1. Three models: $p = 0.1$, $p = 0.3$, and $p = 0.6$. In the first three panels both galaxies have the disc in the x-y plane. In the last two panels we have discs at different angle, first the collision angle was 30°and for the last one 45°.

[5] Gabbasov R F 2006 *Numerical simulation of bars in interacting galaxies* Doctoral thesis (Toluca, México: Universidad Autónoma del Estado de México)
[6] Hohl F 1971 ApJ **168** 343
[7] Sellwood J A 1981 A&A **99** 362
[8] Sellwood J A and Carlberg R G 1984 ApJ **282** 61
[9] Sellwood J A and Athanassoula E 1986 MNRAS **221** 195
[10] Athanassoula E and Sellwood J A 1986 MNRAS **221** 213
[11] Weinberg M D 1985 MNRAS **213** 451
[12] Debattista V P and Sellwood J A 2000 ApJ **543** 704
[13] Weinberg M D and Katz N 2002 ApJ **580** 627
[14] Nogushi M 1987 MNRAS 228 635
[15] Salo H 1990 A&A **243** 118
[16] Gabbasov R F, Rodríguez-Meza M A, Cervantes-Cota J L and Klapp J 2006 Astron. & Astrophys. **449** 1043
[17] Hernquist L 1990 ApJ **356** 359
[18] Rodríguez-Meza M A and Cervantes-Cota J L 2004 MNRAS **350** 671

X Workshop of the Gravitation and Mathematical Physics Division, Mexican Physical Society IOP Publishing
Journal of Physics: Conference Series **545** (2014) 012009 doi:10.1088/1742-6596/545/1/012009

Study of symmetry breaking of charged scalar field: Hydrodynamic version

T Matos[1] and M A Rodríguez-Meza[2]

[1]Departamento de Física, Centro de Investigación y de Estudios Avanzados del IPN, A.P. 14-740, 07000 Mexico City, México
[2]Departamento de Física, Instituto Nacional de Investigaciones Nucleares, Apdo. Postal 18-1027, México, D.F., 11801, México

E-mail: [1]tmatos@fis.cinvestav.mx; [2]marioalberto.rodriguez@inin.gob.mx

Abstract. We rewrite the Klein-Gordon (KG) equation for a complex scalar field as a new Gross-Pitaevskii (GP)-like equation. The potential of the scalar field is a mexican-hat potential and the field is in a thermal bath with one loop contribution. We interpret the new GP equation as a finite temperature generalization of the GP equation for a charged field. We find its hydrodynamic version as well and using it, we derive the corresponding thermodynamics. We also obtain a generalized first law for a charged Bose-Einstein Condensate (BEC).

1. Introduction

Now a days, several astronomical observations give evidence for the existence of the so called dark matter, occupying approximately the 22% of the universe, the rest is contained in another dark component called the dark energy with about 68%, usually associated to a fundamental cosmological constant and the baryonic matter with less than 5%. The standard model to study the universe and its evolution is a model based on assuming that dark matter is a cold component with an equation of state of dust type, and a cosmological constant (Λ) for the other dark component (see for example [1] for a review). This model is usually referred as ΛCDM and is able to fit almost all the observations of the universe we have till now at cosmological scales with a very good accuracy. However, this model has some issues at small scales, like, among others, its prediction of cuspy halos of galaxies. This model predicts that the dark matter at the center of the galaxies must have a cusp density profile, while observations tell us that galaxies prefer to have almost constant density profiles in the center, specially dwarf or low surface brightness galaxies. Alternativelly, a scalar field has been proposed as a dark matter model (see [2, 3] or [4] for a review). In this model the nature of dark matter is a fundamental scalar field, where the predictions at cosmological scales coincide with the ones of the ΛCDM model, but it predicts a flat central density profile as observed in galaxies. In short, in the theoretical Lagrangian, we add two terms, a kinetic term and a potential term for the scalar field. We then fix the form of the potential and with the resultant equations we try to explain some observed astronomical data, like rotation curves of spiral galaxies. In particular, this scalar field model predicts that the galaxy dark halo is core-type. We refer the reader to [4] to see more details and comparisons with observations and with the ΛCDM model.

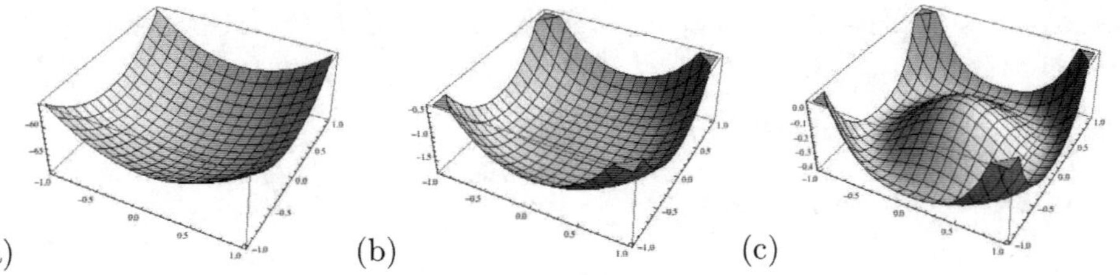

(a) (b) (c)

Figure 1. Potential $V(\Phi)$ as function of Φ. (a) $T = 5$; (b) $T = 2$; and (c) $T = 1$. The plot shows the potential V on the z-axis in terms of the real and imaginary part of the scalar field Φ.

In this work we generalized the model by considering that the scalar field is a complex field and also we consider an electromagnetic term in the Lagrangian. With this model we derive the hydrodynamical version where the potential exhibits a symmetry breaking (SB) and a possible thermodynamics (see also [5] for the real case). Symmetry breaking is normally associated to phase transitions in other areas of physics, therefore its importance, in this work we would like to see whether the symmetry breakdown of the original Lagrangian symmetry has some effect in the structure formation of the universe.

We organise our work in the following form: In the next section we present a general Lagrangian for complex scalar field which includes a term with electromagnetic fields and show how the symmetry appears and how can be broken. Next we show how the Klein-Gordon equation transforms in a generalized Gross-Pitaevskii equation. Three sections follow where we analyze the hydrodynamical version of the generalized Gross-Pitaevskii equation, its Newtonian limit and the thermodynamics of the system. In the final section we write our conclusions of this work.

2. General Lagrangian of the charged scalar field and the gauge symmetry breaking

The general Lagrangian model for a charged scalar field (SF) having a local U(1) symmetry is,

$$\mathcal{L} = (\nabla_\mu \Phi + \mathrm{i}eA_\mu \Phi)(\nabla^\mu \Phi^* - \mathrm{i}eA^\mu \Phi^*) + V(\Phi\Phi^*) - \frac{1}{4}F^{\mu\nu}F_{\mu\nu} \tag{1}$$

where V is the scalar field potential which is a double-well interacting Mexican-hat potential for a complex SF $\Phi(\mathbf{x}, t)$. The scalar field is in thermal equilibrium with a reservoir at temperature T. The thermal interaction between the scalar field and the thermal bath is computed up to one loop of correction, and is given by [5, 6],

$$V(\Phi\Phi^*) = -m^2\Phi\Phi^* + \frac{\lambda}{2}(\Phi\Phi^*)^2 + \frac{\lambda}{4}T^2\Phi\Phi^* - \frac{\pi^2}{90}T^4, \tag{2}$$

This result includes both quantum and thermal contributions. Parameter m is the mass of the scalar field, λ gives us the scalar field self interaction and will be related to a "pressure", in the sense that there is an equation that relates both $p \propto \rho$. In figure 1 we show the potential as function of the complex scalar field (where we consider $m = 1$ and $\lambda = 1$). Units are such that $\hbar = 1$, $c = 1$, and $k_B = 1$.

The dynamics of a SF is governed by the Klein-Gordon (KG) equation, which comes by looking for the extremum of the action corresponding to the Lagrangian (1), it is the equation of motion of a field composed of spinless particles,

$$\Box_E^2 \Phi + \frac{\partial V}{\partial \Phi^*} - 2m^2\phi\Phi = 0, \tag{3}$$

where we have added a first order self-interaction potential ϕ to the SF. For a charged field the D'Alambertian operator is given by,

$$\Box^2_E \equiv (\nabla_\mu + \mathrm{i}eA_\mu)(\nabla^\mu + \mathrm{i}eA^\mu) \tag{4}$$

where $A_\mu = (\mathbf{A}, \varphi)$ is the electromagnetic four potential. Observe that we can rewrite the D'Alambertian as

$$\Box^2_E = (\nabla + \mathrm{i}e\mathbf{A}) \cdot \nabla - (\frac{\partial}{\partial t} + \mathrm{i}e\varphi)\frac{\partial}{\partial t} + \mathrm{i}e\nabla_\mu A^\mu - e^2 A_\mu A^\mu \tag{5}$$

In what follows we will use the Lorentz gauge $\nabla_\mu A^\mu = 0$. It is convenient to consider the total potential V_T by adding to the potential V, the self-interaction contribution and the term $e^2 A_\mu A^\mu = e^2 A^2$, such that

$$\begin{aligned} V_T(\Phi\Phi^*) &= -m^2\Phi\Phi^* + \frac{\lambda}{4}T^2\Phi\Phi^* - e^2A^2\Phi\Phi^* \\ &\quad + \frac{\lambda}{2}(\Phi\Phi^*)^2 - \frac{\pi^2}{90}T^4 - 2m^2\phi\,\Phi\Phi^* \end{aligned} \tag{6}$$

The KG equation now can be written as

$$\Box^2\Phi + \frac{\partial V_T}{\partial \Phi^*} = 0, \tag{7}$$

where now $\Box^2 = \nabla_\mu \nabla^\mu$. Thus we can define an effective mass by

$$m_{eff} = \sqrt{m^2 + e^2A^2}. \tag{8}$$

where the term coming from the electromagnetic potential term is the Proca mass.

Potential (2) has a minimum at $\Phi = 0$ when the temperature $T > T_c$. If $T < T_c$, the point $\Phi = 0$ becomes a maximum and potential (2) has a circular zone of minima with radius

$$\Phi_{min} = \frac{1}{2}\sqrt{T_c^2 - T^2} \tag{9}$$

The critical temperature where the minimum of the potential $\Phi = 0$ becomes a maximum and at which the symmetry is broken is (see figure 1),

$$T_c = \frac{2}{\sqrt{\lambda}}\sqrt{m^2_{eff} + 2m^2\phi}. \tag{10}$$

This temperature is the point where the original $SU(1)$ symmetry of the Lagrangian is broken, because after this point the minimum of the potential is Φ_{min}, where the Lagrangian does not contain the $SU(1)$ symmetry anymore. See [7] for details on BEC and more about this critical temperature.

3. The generalized Gross-Pitaevskii equation

Now for the SF we perform the transformation

$$\Phi = \Psi\, \mathrm{e}^{-\mathrm{i}\hat{m}ct},$$

where $\hat{m}^2 = m^2c^2/\hbar^2$. We have returned to normal units, except that temperature is in energy units ($k_B = 1$).

Journal of Physics: Conference Series **545** (2014) 012009 doi:10.1088/1742-6596/545/1/012009

In terms of function Ψ, the KG equation (3) now reads,

$$
\begin{aligned}
i\hbar\dot{\Psi} + \frac{\hbar^2}{2m}\Box^2\Psi &+ \frac{\lambda}{2mc^2}|\Psi|^2\Psi - mc^2\phi\Psi + ec\varphi\Psi \\
&+ \frac{\lambda T^2}{8mc^2}\Psi = 0,
\end{aligned}
\tag{11}
$$

the complex conjugate equation can be described in the same way. The notation used is: $\dot{\Psi} = \partial\Psi/\partial t$.

Equation (11) is the KG equation (3) or (7) rewritten in terms of the function Ψ and temperature T. This equation is an exact equation defining the field $\Psi(\mathbf{x}, t)$, where ϕ is the external potential acting on the system and the terms in λ represent the interaction potential within the system. When $T \to 0$ and in the non-relativistic limit, that means, in the limit where we take $c \gg 1$, the D' Alambertian $\Box^2 \to \nabla^2$, eq. (11) becomes the Gross-Pitaevskii equation for Bose-Einstein Condensates (BEC) (see for example [9]). And also the static limit of equation (11) is the known Ginzburg-Landau equation. Therefore we will consider equation (11) as a generalization of the Gross-Pitaevskii equation that describe a complex charged scalar field in a thermal bath at finite temperature[7].

4. The Hydrodynamical version

In what follows we transform the generalized Gross-Pitaevskii equation (11) into its analogous hydrodynamical version, for this purpose the ensemble wave function Ψ will be represented in terms of a modulus n and a phase S as,

$$
\Psi = \sqrt{n}\, e^{iS}. \tag{12}
$$

where the phase $S(\mathbf{x}, t)$ is taken as a real function. As usual this phase will define the velocity. We can interpret $n(\mathbf{x}, t) = \rho/M_T$ as the ratio between the number density of particles in the condensed state, $\rho = mn_0 = mN_0/L^3$, being N_0 the number of particles in condensed state and M_T the total mass of the particles in the system. Both, S and n, are functions of time and position (See [8] for a more detailed discussion on this hydrodynamical approach). The concept of SB is often used as a sufficient condition for BEC [7].

So, as can be seen in figure 1, we have that the SF can oscillate around the $\Phi = 0$ minimum. Below the critical temperature T_c, SF oscillates close to the "ring" minimum zone (For a real scalar field we will have that the oscillation ocurrs around the minimum values $\Phi^2_{min} = (T_c^2 - T^2)/4$, and the density will oscillate around $n = \kappa^2(T_c^2 - T^2)/4$ as can be seen by equation (9)), where κ is a scale constant, that will determined in some experiment.

Then, to obtain the hydrodynamical equations, we perform the Madelung transformation (12) in the generalized Gross-Pitaevskii equation (11). We obtain:

$$
\dot{n} + \nabla\cdot(n\mathbf{v}) - \frac{\hbar}{2me}\dot{j} = 0, \tag{13a}
$$

$$
\begin{aligned}
\dot{\mathbf{v}} + (\mathbf{v}\cdot\nabla)\mathbf{v} - \frac{ce}{m}(\mathbf{E} + \mathbf{v}\times\mathbf{B}) &= \\
-c^2\nabla\phi + \frac{\lambda}{m^2c^2\kappa^2}\nabla n &+ \frac{\hbar^2}{m^2}\nabla\left(\frac{\nabla^2\sqrt{n}}{\sqrt{n}}\right) \\
+\frac{\hbar}{2emc^2 n}\mathbf{v}\dot{j} - \frac{\hbar^2}{m^2}\nabla\left(\frac{\partial_t^2\sqrt{n}}{\sqrt{n}}\right) &+ \frac{\lambda}{4m^2}T\nabla T
\end{aligned}
\tag{13b}
$$

where $\mathbf{E}$ and $\mathbf{B} = \nabla\times\mathbf{A}$ are the electric and the magnetic field vectors, respectively. Notice that if we want to compare these equations with the hydrodynamical ones, there are extra terms corresponding with the relativistic character of the Klein-Gordon equation. We have also defined the fluxes

$$\mathbf{j} = 2en(\nabla S + \frac{e}{\hbar}\mathbf{A}), \quad j = 2en(\dot{S} + \frac{e}{\hbar}\varphi), \quad j_\mu = (\mathbf{j}, j) \tag{14a}$$

and

$$\mathbf{v} \equiv \frac{\hbar}{m}(\nabla S + e\mathbf{A}) \tag{15}$$

Notice that in (13b) $\hbar$ enters on the right-hand side through the term containing the gradient of n. This term is usually called the "quantum pressure" and is a direct consequence of the Heisenberg uncertainty principle, it reveals the importance of quantum effects in interacting gases.

Multiplying by n, (13b) can be written as:

$$n\dot{\mathbf{v}} + n(\mathbf{v}\cdot\nabla)\mathbf{v} = n\mathbf{F}_E + n\mathbf{F}_\phi - \nabla p + n\mathbf{F}_Q + \nabla\sigma, \tag{16}$$

where $\mathbf{F}_E = \frac{e}{m}(\mathbf{E} + \mathbf{v}\times\mathbf{B})$ is the electromagnetic force, $\mathbf{F}_\phi = -\nabla\phi$ is the force associated to the external potential ϕ, p can be seen as the pressure of the SF gas that satisfies the equation of state $p = wn^2$, ∇p are forces due to the gradients of pressure and $w = -\lambda/4m^2$ is an interaction parameter, $\mathbf{F}_Q = -\nabla U_Q$ is the quantum force associated to the quantum potential,

$$U_Q = -\frac{\hbar^2}{2m^2}\left(\frac{\nabla^2\sqrt{n}}{\sqrt{n}}\right), \tag{17}$$

and $\nabla\sigma$ is defined as

$$\begin{aligned}\nabla\sigma = \frac{\hbar}{2me}\dot{\mathbf{v}}j \quad &+ \quad \frac{1}{4}\frac{\lambda}{m^2}nT\nabla T \\ &+ \quad \zeta\nabla(\dot{\ln n}) - \frac{\hbar^2 n}{4m^2}\nabla\left(\frac{\ddot{n}}{n}\right),\end{aligned} \tag{18}$$

where the coefficient ζ is given by

$$\zeta = \frac{\hbar^2}{4m^2}\left[-\nabla\cdot(n\mathbf{v}) + \frac{\hbar}{2me}j\right],$$

and the term $\nabla(\dot{\ln n})$ can be written as

$$\nabla(\dot{\ln n}) = -\nabla(\nabla\cdot\mathbf{v}) - \nabla[\nabla(\ln n)\cdot\mathbf{v}] + \frac{1}{m}\nabla[\frac{1}{n}\dot{j}]$$

System (13) is the hydrodynamical representation to equation (11) and is completely equivalent to it.

5. The Newtonian limit

The Newtonian limit is characterized because we expand the time derivatives in order of a small parameter $1/\epsilon^2$, while the space derivatives are expanded in the same parameter as $1/\epsilon$. Neglecting second order time derivatives (terms with $1/\epsilon^4$ and beyond) and products of time derivatives we can simplify system (13) (observe that the velocity $\mathbf{v}$ and $\hbar$ are still second order). In this limit we arrive to the non-relativistic system of equations (13),

$$\dot{n} + \nabla\cdot(n\mathbf{v}) = 0, \tag{19a}$$

$$n\dot{\mathbf{v}} + n(\mathbf{v}\cdot\nabla)\mathbf{v} = n\mathbf{F}_E + n\mathbf{F}_\phi - \nabla p + n\mathbf{F}_Q + \nabla\sigma. \tag{19b}$$

Journal of Physics: Conference Series **545** (2014) 012009 doi:10.1088/1742-6596/545/1/012009

Equation (19a) is the continuity equation, and (19b) is the equation for the momentum transfer. Observe that this last one contains forces due to the external potential, to the gradient of the pressure, viscous forces due to the interactions of the condensate and forces due to the quantum nature of the equations. Quantity $\nabla(\dot{\ln n})$ plays a very important roll, in this limit it reads

$$\nabla(\dot{\ln n}) = -\nabla(\nabla \cdot \mathbf{v}) - \nabla[\nabla(\ln n) \cdot \mathbf{v}].$$

Thus

$$\nabla\sigma = \frac{1}{4}\frac{\lambda}{m} nT\nabla T - \zeta\left[\nabla(\nabla \cdot \mathbf{v}) + \nabla[\nabla(\ln n) \cdot \mathbf{v}]\right], \tag{20}$$

where now we have

$$\zeta = -\frac{\hbar^2}{4m^2}\nabla \cdot (n\mathbf{v}),$$

We interpret the function $\nabla\sigma$ as the viscosity of the system, it contains terms which are gradients of the temperature and of the divergence of the velocity and density (dissipative contributions). The measurement of the temperature dependence in this thermodynamical quantity at the phase transitions might reveal important information about the behavior of the gas due to particle interaction.

6. The Thermodynamics

In what follows we will derive the thermodynamical equations from the hydrodynamical representation. We can derive a conservation equation for a function α, starting with the relationship [5]

$$(n\alpha)^{\cdot} = n\dot{\alpha} + \alpha\dot{n} \tag{21}$$

where α can take the values of ϕ and U_Q, both of them fulfil equation (21). Using the continuity equation

$$\dot{n} + \nabla \cdot (n\mathbf{v}) = 0 \tag{22}$$

in (21) we obtain,

$$(n\alpha)^{\cdot} + \nabla \cdot (n\mathbf{v}\alpha) = -n\mathbf{v} \cdot \mathbf{F}_\alpha + n\dot{\alpha}.$$

where $\mathbf{F}_\alpha = -\nabla\alpha$. The treatment of σ using the above this procedure is more difficult because in general we do not know it explicitly.

As we know, in general (for non-relativistic systems), the total energy density of the system ϵ is the sum of the kinetic, potential and internal energies in this case we have an extra term U_Q due to the quantum potential,

$$\epsilon = ne = \frac{1}{2}nv^2 + n\phi + nu + nU_Q + \psi_E \tag{23}$$

being u the internal energy of the system and

$$\psi_E = \frac{e}{m}(\varphi - \mathbf{v} \cdot \mathbf{A}) \tag{24}$$

the electromagnetic energy potential, defined in terms of the vector potential $\mathbf{A}$ and the electric potential φ.

Then from (23) we have that u will satisfy the equation

$$(nu)^{\cdot} + \nabla \cdot \mathbf{J}_u - \nabla \cdot \mathbf{J}_\rho + n\dot{\phi} = -p\nabla \cdot \mathbf{v}, \tag{25}$$

being $\mathbf{J}_u$ the energy current, given by a energy flux and a heat flux, $\mathbf{J}_q$,

$$\mathbf{J}_u = nu\mathbf{v} + \mathbf{J}_q + \mathbf{J}_B - p\mathbf{v},$$

where $\nabla \cdot \mathbf{J}_q = \mathbf{v} \cdot (\nabla\sigma)$, and $\nabla \cdot \mathbf{J}_B = \mathbf{v} \cdot (n\mathbf{j}_B)$, being $\mathbf{j}_B$ given by the continuity equation of the vector potential $\mathbf{A}$

$$\frac{\partial \mathbf{A}}{\partial t} + (\mathbf{v} \cdot \nabla)\mathbf{A} = -(\mathbf{A} \cdot \nabla)\mathbf{v} + \frac{m}{e}\mathbf{j}_B, \tag{26}$$

expressions that as we can see is related in a direct way to the velocity and gradients of temperature in the condensate, and is the one that shows in an explicit way the temperature dependence of the thermodynamical equations. With these definitions at hand we have,

$$(nu)^{\cdot} + \nabla \cdot (n\mathbf{v}u + \mathbf{J}_q + \mathbf{J}_B - p\mathbf{v} - \mathbf{J}_\rho) + n\dot{\phi} = -p\nabla \cdot \mathbf{v}. \tag{27}$$

where $\mathbf{J}_\rho = n\mathbf{v}_\rho = \hbar^2/4m^2(\nabla \ln n)^{\cdot}$. In order to find the thermodynamical quantities of the system in equilibrium (taking p as constant on a volume L), we restrict the system to the regime where the auto-interacting potential is constant in time, with this conditions at hand for (27) we have:

$$(nu)^{\cdot} + \nabla \cdot (n\mathbf{v}u + \mathbf{J}_q + \mathbf{J}_B - p\mathbf{v} - \mathbf{J}_\rho) = -p\nabla \cdot \mathbf{v} \tag{28}$$

As always the first term will represent the change in the internal energy of the system, $-p\nabla \cdot \mathbf{v}$ is the work done by the pressure and $\nabla \cdot \mathbf{v}$ is related to the change in the volume. The term with $\mathbf{J}_q$ contains terms related to the heat generated by gradients of the temperature ∇T and dissipative forces due to viscous forces $\sim \nabla(\nabla \cdot \mathbf{v})$. And finally but most important we have an extra term, $\nabla \cdot \mathbf{J}_\rho$, due to gradients of the quantum potential (17).

Integrating this expression on a close region, we obtain

$$\begin{aligned}\frac{\mathrm{d}}{\mathrm{d}t}\int nu\,\mathrm{d}V + \oint (\mathbf{J}_q + \mathbf{J}_B + p\mathbf{v}) \cdot \mathbf{n}\,\mathrm{d}S &- \oint \mathbf{J}_\rho \cdot \mathbf{n}\,\mathrm{d}S \\ &= -p\frac{\mathrm{d}}{\mathrm{d}t}\int \mathrm{d}V.\end{aligned}$$

Equation (28) is the continuity equation for the internal energy of the system and as usual, from here we have an expression that would describe the thermodynamics of the system in an analogous way as does the first law of thermodynamics, in this case for the KG equation or a BEC. This reads

$$\mathrm{d}U = \hat{\mathrm{d}}Q + \hat{\mathrm{d}}Q_B + \hat{\mathrm{d}}A_Q - p\mathrm{d}V \tag{29}$$

where $U = \int nu\,\mathrm{d}V$ is the internal energy of the system, and as we can see, its change is due to a combination of heat Q added to the system and work done on the system (pressure dependent), and

$$\frac{\hat{\mathrm{d}}A_Q}{\mathrm{d}t} = \frac{\hbar^2}{4m^2}\oint n(\nabla \ln n)^{\cdot} \cdot \mathbf{n}\ \mathrm{d}S = \oint n\mathbf{v}_\rho \cdot \mathbf{n}\,\mathrm{d}S,$$

is the corresponding quantum heat flux due to the quantum nature of the KG equation.

The first and third terms on the right hand side of equation (29) would make the crucial difference between a classical and a quantum first law of thermodynamics. Analogously, for the magnetic heat we have

$$\begin{aligned}\frac{\hat{\mathrm{d}}Q_B}{\mathrm{d}t} &= \int \nabla \cdot \mathbf{J}_B\,dV = \int \mathbf{v} \cdot (n\mathbf{j}_B)\,dV \\ &= \frac{m}{e}\int n\left(\frac{\partial \mathbf{A}}{\partial t} + (\mathbf{v} \cdot \nabla)\mathbf{A} + (\mathbf{A} \cdot \nabla)\mathbf{v}\right) \cdot \mathbf{v}\,dV\end{aligned} \tag{30}$$

where the vector potential **A** fulfills the Maxwell equations, in terms of the fluxes (14) it reads

$$F^{\mu\nu}{}_{,\nu} = -j_\nu \tag{31}$$

where as usual $F_{\mu\nu} = \partial_\mu A_\nu - \partial_\nu A_\mu$.

To complete the description we write the Maxwell equations. In terms of the vector and the electric potential, the Maxwell equations are given by

$$\Box\mathbf{A} = -\frac{1}{\kappa^2}\mathbf{j} \tag{32a}$$

$$\Box\varphi = -\frac{1}{\kappa^2}j \tag{32b}$$

where we have used the Lorentz gauge. Observe that the fluxes contain the information of the velocity of the fluid and of the electromagnetic term as well.

7. Conclusions

In this work we have studied the $U(1)$ SB of the Klein-Gordon Lagrangian with temperature contributions to the effective Mexican hat potential of a system of weakly interacting bosons. We did this using a simple transformation in order to rewrite the Klein-Gordon equation as a GP-like one. This is not surprising, because the Klein-Gordon equation represents bosons with spin zero and the GP equation represents bosons in condensed state. We interpret this GP-like equation as a generalization of the normal GP one but for relativistic particles and finite temperature. Using the Madelung transformation we rewrite again the Klein-Gordon equations in their hydrodynamical version where we have obtained several thermodynamical relations for the Gross-Pitaevskii equation (11). With this thermodynamical version it is much simpler to interpret the physics of the phenomena happening during the SB. For example we find that in certain conditions the system becomes super-conductor and under other conditions it becomes super-fluid. In this interpretations super-conductivity and super-fluidity can be understood from first principles of quantum field theory in a unified way.

References

[1] Liddle A 2003 *An Introduction to Modern Cosmology* Second edition (London: Wiley)
[2] Guzmán F S, Matos T and Villegas H 1999 Astron. Nachr. **320** 97-104
[3] Matos T and Guzmán F S 2000 Class. Quant. Grav. **17** L9-L16. *Preprint* arXiv:gr-qc/9810028
[4] Suárez A, Robles V H and Matos T 2013 *Preprint* arXiv:1302.0903
[5] Matos T and Suárez A 2011 EPL **96** 56005
[6] Kolb E W and Turner M S 1990 *The Early Universe* (New York: Addison-Wesley Publishing Co.)
[7] Castellanos E and Matos T 2012 *Preprint* arXiv:1202.3793
[8] Rodríguez-Meza M A, Hernández-Almada and Matos T 2012 Rev. Mex. Fis. E **58** 53
[9] F. S. Guzman and L. A. Urena-Lopez, Phys. Rev. D **69** (2004) 124033 [gr-qc/0404014].

X Workshop of the Gravitation and Mathematical Physics Division, Mexican Physical Society IOP Publishing
Journal of Physics: Conference Series **545** (2014) 012010 doi:10.1088/1742-6596/545/1/012010

On gravitational waves generated during inflation

Claudia Moreno, José Edgar Madriz Aguilar

Departamento de Matemáticas, Centro Universitario de Ciencias Exactas e ingenierías (CUCEI), Universidad de Guadalajara (UdG), Av. Revolución 1500 S.R. 44430, Guadalajara, Jalisco, México.

E-mail: claudia.moreno@cucei.udg.mx, edgar.madriz@red.cucei.udg.mx

Abstract. We study cosmological gravitational waves generated during inflation under the influence of a decaying cosmological "constant". A non-perturbative contribution of the dynamical cosmological "constant" to the tensor modes is considered. As applications of the model we study the well-known cases $\Lambda(t) = \sigma H^2$ and $\Lambda(t) = \vartheta H$. The spectrum of gravitational waves results scale invariant only for the first case, leaving this to new proposals of the form: $\Lambda(t) = f(H, H^2)$, in order to include inflation in some $\Lambda(t)$CDM models. We also found that the non-perturbative contributions of $\Lambda(t)$, accelerate the decreasing of the amplitude of gravitational waves during a power-law inflationary stage, by an exponential factor.

1. Introduction

The relic background of gravitational waves is a prediction of inflationary cosmology [1, 2]. Their amplitude is related with the energy scale of inflation and the are observationally detectable by studies of B-mode polarization in the cosmic microwave background radiation (CMBR), if the energy of inflation is larger than $\sim 3 \cdot 10^{15}$ GeV [3, 4, 5]. The technical mechanism for generating primordial gravitational waves has mainly two branches: the coordinate approach of Bardeen and the covariant formalism [6, 7, 8].

On the other hand, the present acceleration in the expansion of the universe is now accepted as an observational fact [9, 10]. The Λ-CDM model [11, 12, 13] is a concordance model that fits observations satisfactorily but unfortunately gives no more profound explanations as for instead on the nature or origin of the cosmological constant. Moreover, it still suffers from the cosmological constant and the cosmic coincidence problems [14].

A different proposal to alleviate these problems are the $\Lambda(t)$-CDM models [15]. In these models the dynamical cosmological constant represents a decaying vacuum energy capable to generate the observed acceleration in the cosmic expansion, without fall in contradictions with nucleosynthesis and predictions of structure formation. According to some models of QCD chiral phase transitions during the early universe, it is generated a vacuum energy density that scales as $\Lambda \simeq m^3 H$, with $m \simeq 150$ MeV being the energy scale of QCD vacuum transitions and H denoting the Hubble parameter [15, 16, 17, 18, 19, 20, 21, 22]. Taking these results into account, two of the most employed models for $\Lambda(t)$ are $\Lambda(t) \simeq H^2$ and $\Lambda(t) \simeq m^3 H$ [23]. A cosmological scenario derived from $\Lambda(t) \simeq m^3 H$ has been sutied by S. Carneiro, J. S. Alcaniz and collaborators, establishing its observational viability starting from a radiation dominated epoch and

Journal of Physics: Conference Series **545** (2014) 012010 doi:10.1088/1742-6596/545/1/012010

finishing in the present epoch [23]. However, their results do not consider the inflationary epoch.

The present lecture notes are based on our paper [24]. In this proceeding notes we study the generation of the relic background of gravitational waves during inflation under the presence of a dynamical cosmological constant. We consider a new approach in which a non-perturbative dependence of $\Lambda(t)$ in the tensor modes is taken. We work in the transverse-traceless (TT) gauge and consider two applications of the model.

2. Basics on $\Lambda(t)$-CDM Model

The Einstein field equations with cosmological constant read

$$R_{\mu\nu} - \frac{1}{2}Rg_{\mu\nu} + \Lambda g_{\mu\nu} = 8\pi G T_{\mu\nu}, \tag{1}$$

where $R_{\mu\nu}$ is the Ricci tensor, R is the Ricci scalar and Λ is the cosmological constant. As it is well-known due to Bianchi identities $\nabla_\mu G^{\mu\nu} = 0$, and the Riemannian metricity condition $\nabla_\mu g_{\alpha\beta} = 0$, the Eq. (1) implies the conservation of the energy momentum tensor $\nabla^\mu T_{\mu\nu} = 0$. In the case of $\Lambda(t)$-CDM models the dynamical cosmological constant $\Lambda(t)$ is added on the right hand side of Einstein equations in the form [15]

$$R_{\mu\nu} - \frac{1}{2}Rg_{\mu\nu} = \Lambda(t) g_{\mu\nu} + 8\pi G T_{\mu\nu}, \tag{2}$$

and this agrees with the idea that $\Lambda(t)$ represents a dynamical (decaying) vacuum energy. Thus on a Friedmann-Robertson-Walker (FRW) metrical background the Friedmann equation and the conservation equation derived from (2) as a consequence of Bianchi identities are

$$H^2 = \frac{8\pi G}{3}\rho + \frac{\Lambda}{3}, \tag{3}$$

$$\dot{\rho} + 3H(\rho + p) = -\frac{\dot{\Lambda}}{8\pi G}, \tag{4}$$

where $H = \dot{a}/a$ is the Hubble parameter, the dot means time derivative. The equation (4) is like the standard continuity equation for a perfect fluid on cosmological settings, but with a source term $-\frac{\dot{\Lambda}}{8\pi G}$. However, this equation can be also written in the standard form if we define an effective energy-momentum tensor for a perfect fluid with an effective energy density $\rho_{eff} = \rho + (\Lambda/8\pi G)$ and an effective pressure $p_{eff} = p - (\Lambda/8\pi G)$.

3. Relic background of gravitational waves from $\Lambda(t)$-CDM model: the formalism

In the TT-gauge, the perturbed line element for a spatially flat FRW spacetime can be written in the form

$$ds^2_{pert} = dt^2 - a(t)^2(\delta_{ij} + \mathcal{Q}_{ij})dx^i dx^j, \tag{5}$$

where the tensor modes $\mathcal{Q}_{ij}(t,\vec{r})$ satisfy: $tr(\mathcal{Q}_{ij}) = \mathcal{Q}^i{}_i = 0$ and $\mathcal{Q}^i{}_{j;i} = 0$. The 3D spatial components of the metric can be written as $g_{ij} = -a^2(t)(\delta_{ij} + \mathcal{Q}_{ij})$ whereas its 3D contravariant components can be linearly approximated by $g^{ij} \simeq -a^{-2}(t)(\delta^{ij} - \mathcal{Q}^{ij})$.

Due to the well-known fact that in the TT-gauge only space-space components contribute for tensor fluctuations, the dynamics of the tensor modes is in this case given by linearized Einstein equations free of sources

$$\delta R_{ij} - \frac{1}{2}\delta R\, g^{(b)}_{ij} - \left(\frac{1}{2}R^{(b)} - \Lambda(t)\right)\mathcal{Q}_{ij} = 0, \tag{6}$$

where the label (b) denotes quantities computed with the background metric derived from (5), δR_{ij} is the linearized Ricci tensor and δR is the Ricci scalar up to first order in $\mathcal{Q}_{ij}$.

By using (5), after some algebra the equations (6) can be put in the form

$$\ddot{\mathcal{Q}}^i{}_j + 3H\dot{\mathcal{Q}}^i{}_j - \frac{1}{a^2}\nabla^2\mathcal{Q}^i{}_j + \Lambda\mathcal{Q}^i{}_j = 0, \tag{7}$$

where the dot is denoting derivative with respect the cosmic time. In order to study the impact of a dynamical cosmological constant on the background of gravitational waves during inflation, we must obtain the exact dependence of the tensor of metric fluctuations $\mathcal{Q}_{ij}$, with $\Lambda(t)$. Thus, it is natural from (7) to expect the dependence: $\mathcal{Q}_{ij} = \mathcal{Q}_{ij}(t, \vec{r}, \Lambda(t))$, that we will also express as $\mathcal{Q}_{ij}(\bar{x}, \Lambda(t))$. From now on, by simplicity in the notation, we will omit the time dependence of $\Lambda(t)$ in some equations and we will simply write Λ.

Now, we assume that the exact solution of (7) can be written as a power series in $\Lambda(t)$

$$\mathcal{Q}_{ij} = \sum_{l=0}^{\infty} h_{ij}^{(l)}\Lambda^l = h_{ij}^{(0)} + h_{ij}^{(1)}\Lambda + h_{ij}^{(2)}\Lambda^2 + \mathcal{O}(\Lambda^3), \tag{8}$$

where $h_{ij}^{(l)} = (1/l!)(\partial^l\mathcal{Q}_{ij}/\partial\Lambda^l)_{\Lambda=0}$ are the coefficients of the Taylor expansion. If we study separable solutions of the form $\mathcal{Q}_{ij}(\bar{x}, \Lambda) = h_{ij}(\bar{x})\Gamma(\Lambda)$, the taylor coefficients become $h_{ij}^{(l)} = h_{ij}(\bar{x})(1/l!)(d^l\Gamma/d\Lambda^l)_{\Lambda=0}$. Imposing also the condition: $\mathcal{Q}_{ij}(\Lambda = 0) = h_{ij}(\bar{x})$, which is the condition to recover the standard case without Λ when we put $\Lambda = 0$ in (8), the series (8) can be written in the form

$$\mathcal{Q}_{ij} = h_{ij}\left[1 + \widetilde{\alpha}_1\Lambda + \frac{1}{2!}\widetilde{\alpha}_2\Lambda^2 + \mathcal{O}(\Lambda^3)\right]. \tag{9}$$

In this moment, we may content ourselves with the expansion (9) up to linear order in $\Lambda(t)$, as it is done for example in [3]. However, a better approximation can be made if we consider the case $\widetilde{\alpha}_1 = \alpha_0$, $\widetilde{\alpha}_2 = \alpha_0^2$,... On this particular case, the expression (9) yields

$$\mathcal{Q}_{ij}(\bar{x}, \Lambda) = h_{ij}(\bar{x})e^{\alpha_0\Lambda(t)}, \tag{10}$$

where α_0 has units of Λ^{-1}. The approximation (10) is non-perturbative in $\Lambda(t)$ and this may be considered as an advantage. Moreover, this equation expresses the contribution of both the usual tensor metric fluctuations $h_{ij}(\bar{x})$ (i.e. without any Λ) and $\Lambda(t)$, to the general tensor metric fluctuations $\mathcal{Q}_{ij}$, in a separated way. In this letter we will work with the approximation (10).

Now by means of (10), writing the equation (7) in terms of the usual tensor fluctuations $h^i{}_j$ we arrive to

$$\ddot{h}^i_j + (3H + 2\alpha_0\dot{\Lambda})\dot{h}^i_j - \frac{1}{a^2}\nabla^2 h^i_j + \left(\alpha_0\ddot{\Lambda} + \alpha_0^2\dot{\Lambda}^2 + 3\alpha_0 H\dot{\Lambda} + \Lambda\right)h^i_j = 0. \tag{11}$$

Introducing the auxiliary field $\chi^i{}_j(t, \vec{r})$ defined by the transformation

$$h^i{}_j(t, \vec{r}) = e^{-\frac{1}{2}\int(3H + 2\alpha_0\dot{\Lambda})dt}\chi^i{}_j(t, \vec{r}), \tag{12}$$

the equation (11) can be written as

$$\ddot{\chi}^i{}_j - \frac{1}{a^2}\nabla^2\chi^i{}_j - \left(\frac{9}{4}H^2 + \frac{3}{2}\dot{H} - \Lambda\right)\chi^i{}_j = 0. \tag{13}$$

Following the usual quantization process, we implement the Fourier expansion for ${\chi^i}_j$

$$\chi^i_j(t,\vec{r}) = \frac{1}{(2\pi)^{3/2}} \int d^3k \sum_{\alpha=+,\times} {}^{(\alpha)}e^i_j \left[a^{(\alpha)}_k e^{i\vec{k}\cdot\vec{r}} \xi_k(t) + a^{(\alpha)\,\dagger}_k e^{-i\vec{k}\cdot\vec{r}} \xi^*_k(t) \right], \tag{14}$$

with the creation and annihilation operators $a^{(\alpha)\,\dagger}_k$ and $a^{(\alpha)}_k$ respectively, satisfying

$$\begin{aligned} \left[a^{(\alpha)}_k, a^{(\alpha')\,\dagger}_{k'}\right] &= g^{\alpha\alpha'}\delta^{(3)}(\vec{k}-\vec{k}'), \\ \left[a^{(\alpha)}_k, a^{(\alpha')}_{k'}\right] &= \left[a^{(\alpha)\,\dagger}_k, a^{(\alpha')\,\dagger}_{k'}\right] = 0, \end{aligned} \tag{15}$$

and where the following properties for the polarization tensor e_{ij} hold

$$\begin{aligned} {}^{(\alpha)}e_{ij} &= {}^{(\alpha)}e_{ji}, \quad k^i\,{}^{(\alpha)}e_{ij} = 0, \\ {}^{(\alpha)}e_{ii} &= 0, \quad {}^{(\alpha)}e_{ij}[-\vec{k}] = {}^{(\alpha)}e^*_{ij}[\vec{k}]. \end{aligned} \tag{16}$$

We also require the modes $\xi_k(t)$ satisfy the commutation relation

$$\left[\xi_k(t,\vec{r}), \dot{\xi}_k(t,\vec{r}')\right] = i\delta^{(3)}\left(\vec{r}-\vec{r}'\right), \tag{17}$$

which for (14) reduces to

$$\xi_k\dot{\xi}^*_k - \xi^*_k\dot{\xi}_k = i, \tag{18}$$

with the asterisk denoting the complex conjugate operation. When this condition holds true the modes are normalizable on the UV-sector.

In view of (13) and (14), the modes ξ_k obey the dynamical equation

$$\ddot{\xi}_k + \left(\frac{k^2}{a^2} - \frac{9}{4}H^2 - \frac{3}{2}\dot{H} + \Lambda\right)\xi_k = 0. \tag{19}$$

Thus, in principle, for a given $\Lambda(t)$ we can obtain a normalized solution for the k-modes ξ_k, just by solving (19) and imposing the normalization condition (18) to that solution.

Once a normalized solution for the k-modes is obtained, we are in position to calculate the spectrum for tensor fluctuations on Super-Hubble scales i.e. on the IR-sector.

The amplitude of the tensor fluctuations $< h^2 >_{IR} = < 0|{h^i}_j {h_i}^j|0 >$ on the IR sector is given in this case by

$$\langle h^2\rangle_{IR} = \frac{1}{2\pi^2} e^{-\int(3H+2\alpha\dot{\Lambda})dt} \int_0^{\epsilon k_H} \frac{dk}{k} k^3 \left[\xi_k\xi^*_k\right]_{IR}, \tag{20}$$

where $\epsilon = k^{IR}_{max}/k_P \ll 1$ is a dimensionless parameter, being $k^{IR}_{max} = k_H(t_i)$ the wave-number related to the Hubble radius at the time t_i, which is the time when the modes re-enter to the horizon at the end of inflation. The k_P is the Planckian wave-number. For a typical Hubble parameter at the end of inflation $H = 0.5\cdot 10^{-9}M_P$, values of ϵ on the range of 10^{-5} to 10^{-8} correspond to the number of e-foldings $N_e = 63$.

Journal of Physics: Conference Series **545** (2014) 012010 doi:10.1088/1742-6596/545/1/012010

4. Applications

4.1. The case of a decaying $\Lambda(t)$

Let us considering a power-law inflationary stage, thereby the scale factor is taken to be $a(t) = a_0(t/t_0)^p$, with a_0 being the present value of the scale factor, t_0 the present time, and $p > 1$ a constant parameter compatible with an accelerated expansion. The equation for the modes (19) reduces in this case to

$$\ddot{\xi}_k + \left(\frac{k^2}{a_0^2}t^{-2p} - \frac{(9/4)p^2 - (3/2)p}{t^2} - \Lambda(t)\right)\xi_k = 0. \tag{21}$$

For a decaying $\Lambda(t)$ of the form $\Lambda(t) = \varpi/t^2$, with ϖ being a constant parameter, the expression (21) reads

$$\ddot{\xi}_k + \left(\frac{k^2}{a_0^2}t^{-2p} - \frac{(9/4)p^2 - (3/2)p + \varpi}{t^2}\right)\xi_k = 0. \tag{22}$$

The general solution for (22) is given by

$$\xi_k(t) = C_1\Gamma(1+\nu)2^{\nu}t^{\frac{1}{2}-(p-1)\nu}z(t)^{-\nu}\mathcal{J}_{\nu}[z(t)] + C_2\Gamma(1-\nu)2^{-\nu}t^{\frac{1}{2}+(p-1)\nu}z(t)^{\nu}\mathcal{J}_{-\nu}[z(t)], \tag{23}$$

where $\mathcal{J}_\nu$ is the Bessel Function, $z(t) = \frac{k}{a_0(p-1)}(t/t_0)^{1-p}$, and $\nu = \sqrt{1+4\beta}/(2(p-1))$, with $\beta = (9/4)p^2 - (3/2)p + \varpi$. Unfortunately, this solution is not normalizable. A normalizable solution of (22) can be obtained when we write (22) in terms of the conformal time τ. Thus, for the conformal time during inflation $\tau = [a_0/(t_0^p(1-p))]t^{1-p}$ the equation (22) gives

$$\frac{d^2\xi_x}{d\tau^2} - \frac{p}{1-p}\left(\frac{1}{\tau}\right)\frac{d\xi_k}{d\tau} + \left(\kappa^2 - \frac{\kappa_0^2}{\tau^2}\right)\xi_k = 0, \tag{24}$$

where $\kappa = (t_0^{2p}/a_0^2)k$ and $\kappa_0^2 = (1-p)^{-2}[(9/4)p^2 - (3/2)p - \varpi]$. The general solution of this equation is given in terms of the first and second kind Hankel functions $\mathcal{H}_\mu^{(1)}$ and $\mathcal{H}_\mu^{(2)}$ in the form

$$\xi_k(\tau) = A_1\tau^{-\frac{1}{2(p-1)}}\mathcal{H}_\mu^{(1)}[w(\tau)] + A_2\tau^{-\frac{1}{2(p-1)}}\mathcal{H}_\mu^{(2)}[w(\tau)] \tag{25}$$

being $w(\tau) = k\tau$ and $\mu = [1/(2(p-1))]\sqrt{1+4\kappa_0^2(p-1)^2}$. The normalized solution to (24) is then

$$\xi_k(\tau) = i\sqrt{\frac{\pi}{4}}\left[\frac{(p-1)t_0}{a_0^{1/p}}\right]^{-\frac{p}{2(p-1)}}\tau^{-\frac{1}{2(p-1)}}\mathcal{H}_\mu^{(1)}[w(\tau)]. \tag{26}$$

Now, getting back to the cosmic time, the normalized solution (26) reads

$$\xi_k(t) = i\sqrt{\frac{\pi}{4(p-1)}}\,t^{1/2}\mathcal{H}_\mu^{(1)}[w(t)], \tag{27}$$

with $w(t) = [kt_0^p/(a_0(p-1))t^{1-p}]$. In view of (27), the amplitude for gravitational waves on Super-Hubble scales ($k \gg k_H$), according to the expression (20) gives

$$\left\langle h^2\right\rangle_{IR} = \frac{t_0^{(3-2\mu)p}}{8\pi^3}\frac{(2a_0)^{2\mu}}{(p-1)^{1-2\mu}}\Gamma^2(\mu)t^{\gamma}e^{-\frac{2\alpha\varpi}{t^2}}(\epsilon k_H)^{3-2\mu}, \tag{28}$$

where $\gamma = 1 - 3p - 2\mu(1-p)$. Therefore for the spectrum of gravitational waves in this case we obtain

$$\mathcal{P}_g(k) = \frac{t_0^{(3-2\mu)p}}{8\pi^3}\frac{(2a_0)^{2\mu}}{(p-1)^{1-2\mu}}\Gamma^2(\mu)t^{\gamma}e^{-\frac{2\alpha\varpi}{t^2}}k^{3-2\mu}\Bigg|_{k=\epsilon k_H}. \tag{29}$$

Journal of Physics: Conference Series **545** (2014) 012010 doi:10.1088/1742-6596/545/1/012010

Clearly for $\mu \approx 3/2$ the spectrum is nearly scale invariant. The scale invariance is achieved when the condition

$$\frac{1+3p(3p-2)-4\varpi}{9(p-1)^2}=1, \tag{30}$$

is valid. The solution of (30) for ϖ is then $\varpi = 3p-2$. The accelerated expansion during inflation $p>1$ leaves to the restriction: $\varpi > 1$. Hence, a nearly scale invariant spectrum for gravitational waves under the presence of a $\Lambda(t) = \varpi/t^2$ is perfectly possible if we restrict ourselves to $\varpi > 1$.

In the particular case of $\Lambda(t) = \sigma H^2$ we would have $\varpi = \sigma p^2$. Thus the condition (30) reduces to

$$\frac{1+(9-4\sigma)p^2-6p}{9(p-1)^2}=1. \tag{31}$$

Then for σ we have the restriction $\sigma = (3p-2)/p^2$. Hence, for $p>1$ we obtain $0<\sigma \leq 9/8 \simeq 1.125$.

4.2. The case of a Cosmological Constant

In the case of a cosmological constant Λ_0 with a de-Sitter expansion $a(t) = a_0 e^{H_0 t}$, with H_0 being the constant value of the Hubble parameter during inflation, the equation for the modes (19) reads

$$\ddot{\xi}_k + \left(\frac{k^2}{a_0^2}e^{-2H_0 t} - \frac{9}{4}H_0^2 - \Lambda_0\right)\xi_k = 0. \tag{32}$$

This equation has for solution

$$\xi_k(t) = D_1 \mathcal{H}_\lambda^{(1)}[x(t)] + D_2 \mathcal{H}_\lambda^{(2)}[x(t)], \tag{33}$$

being $\mathcal{H}_\lambda^{(1,2)}$ the first and second type Hankel functions, $\lambda = [1/(2H_0)]\sqrt{9H_0^2+4\Lambda_0}$ and $x(t) = [k/(a_0H_0)]e^{-H_0 t}$. Invoking (18) and taking a Bunch-Davies vacuum condition, the normalized solution for (32) is given by

$$\xi_k(t) = \frac{i\sqrt{3\pi}}{2\Lambda_0^{1/4}}\mathcal{H}_\lambda^{(1)}[x(t)]. \tag{34}$$

By using (34) the amplitude for gravitational waves on cosmological scales (20), acquires the form

$$\left\langle h^2\right\rangle_{IR} = \frac{3}{\pi^3}\frac{2^{-3+2\lambda}}{\sqrt{\Lambda_0}}\frac{\Gamma^2(\lambda)}{(a_0H_0)^{-2\lambda}}e^{-(3-2\lambda)H_0 t}(\epsilon k_H)^{3-2\lambda}, \tag{35}$$

and the spectrum for gravitational waves results in this case

$$\mathcal{P}_g(k) = \frac{3}{\pi^3}\frac{2^{-3+2\lambda}}{\sqrt{\Lambda_0}}\frac{\Gamma^2(\lambda)}{(a_0H_0^2)^{-2\lambda}}e^{-(3-2\lambda)H_0 t}k^{3-2\lambda}\Big|_{k=\epsilon k_H}. \tag{36}$$

It can be easily check from (36) that the spectral index for gravitational waves under the presence of a cosmological constant is $n_{gw} = 4-2\lambda$. Thus, in principle, for values of Λ_0 obeying $4\Lambda_0/(9H_0^2) \ll 1$, we may have a nearly scale invariant spectrum $n_{gw} \simeq 1$.

A special case that deserves our attention is when $\Lambda_0 = \vartheta H_0$. As we mentioned before, in this case $H_0 = \vartheta/3$ and $\Lambda_0 = \vartheta^2/3$, and therefore $\lambda = \sqrt{21}/2 \simeq 2.2913$. Unfortunately, for this value of λ the spectral index becomes $n_{gw} \simeq -0.5826$, which is far away from the scale invariance, thus entering in contradiction with observations, which from WMAP7 + BAO + H_0-Mean analysis give $n_{gw} \simeq 0.963 \pm 0.012$ [25].

We may interpret this result for $\Lambda(t) = \vartheta H(t)$, as during inflation, at least with respect to the spectrum of gravitational waves, it seems more convenient to use $\Lambda(t) = \sigma H^2(t)$ than $\Lambda(t) = \vartheta H(t)$. However, the fact that the former time dependence $\Lambda(t) = \vartheta H(t)$ works very well for the radiation-dominated, matter-dominated and the present epochs [23, 26], suggests that to include inflation we might consider $\Lambda(t) = f(H, H^2)$, in such a way that during inflation the dominant term is H^2 and after inflation it becomes sub-dominant in front of H. One example of this kind of time dependence for $\Lambda(t)$ would be $\Lambda(t) = (1 - \eta_{sr})H^2 + \eta_{sr}H$, where η_{sr} is the slow-roll parameter during inflation defined by $\eta_{sr} = -\dot{H}/H^2$. As it is well-known during inflation $\eta_{sr} \ll 1$ while when inflation ends $\eta_{sr} = 1$. Thus, we obtain the desired dominance of H^2 during inflation and of H for the resting epochs.

5. Conclusions

In this lecture notes we have shown the impact of a dynamical cosmological constant over the background of gravitational waves generated during inflation. Our results have some differences with respect to the ones we can find in the literature, for example, we incorporate a nonperturbative contribution of the dynamical cosmological constant $\Lambda(t)$ to the tensor modes (see eq.(10)). The fact that the contribution is non-perturbative in $\Lambda(t)$ allow us mainly, to calculate new contributions of $\Lambda(t)$ on the spectrum and the mean square amplitude $< h^2 >$ on the IR-sector (cosmological scales), at all orders in $\Lambda(t)$. We found that the presence of $\Lambda(t)$ speeds up the decreasing of the gravitational waves amplitude during a power law inflation by an exponential factor.

Two applications were in order. In the case of $\Lambda(t) = \sigma H^2$, for a power law expansion, we obtain a scale invariant gravitational spectrum at the end of inflation for $0 < \sigma < 1.125$. In the case of $\Lambda = \vartheta H$ we obtain a spectral index for gravitational waves $n_{gw} \simeq -0.5826$, which is in contradiction with observational data. Therefore, due to the fact that $\Lambda(t) = \vartheta H$ seems to work very well from the radiation dominated epoch to the present time, one way to extend a $\Lambda(t)$-CDM model to inflation, compatible with observations, would be of the form $\Lambda(t) = f(H, H^2)$. One proposal we can make is $\Lambda(t) = (1 - \eta_{sr})H^2 + \eta_{sr}H$, where η_{sr} is the slow-roll parameter during inflation defined by $\eta_{sr} = -\dot{H}/H^2$. In this manner, $\eta_{sr} \ll 1$ during inflation and $\eta_{sr} = 1$ when it ends, reproducing thus the desired dominance of H^2 during inflation and of H forthe resting periods in the cosmic history.

References

[1] A. Starobinsky, JETP Lett. **30** (1979) 682.
[2] E. Stewart, D. Lyth, Phys. Lett. B **302** (1983) 171.
[3] M. Kamionkowski, A. Kosowsky, A. Stebbins, Phys. Rev. D **55** (1997) 7368.
[4] M. Kamionkowski, A. Kosowsky, A. Stebbins, Phys. Rev. Lett. **78** (1997) 2058.
[5] L. Knox, Y. Song, Phys. Rev. Lett. **89** (2002) 011303.
[6] B. Allen, Phys. Rev. D **37** (1998) 2078.
[7] J. Bardeen, Phys. Rev. D **22** (1980) 1882.
[8] G. F. R. Ellis, M. Bruni, Phys. Rev. D **40** (1989) 1804.
[9] A. G. Riess et al., Astron. J. **116** (1998) 1009.
[10] S. Perlmutter et al., Nature **391** (1998) 51.
[11] L. Ford and L. Parker, Phys. Rev. D **16** (1977) 245.
[12] M. de Garcia, Phys. Rev. D **48** (1993) 647.
[13] A. B. Henriquez, Phys. Rev. D **49** (1994) 1771.
[14] Eric V. Linder, Gen. Relativ. **40** (2008) 329.

[15] J. S. Alcaniz, Braz. J. Phys. **36** (2006) 1109.
[16] R. Schitzhold, Phys. Rev. Lett. **89** (2002) 081302.
[17] F. R. Klinkhamer, G. E. Volovik, Phys. Rev. D **79** (2009) 063527.
[18] F. R. Urban, A. R. Zhitnitsky, Phys. Rev. D **80** (2009) 063001
[19] F. R. Urban, A. R. Zhitnitsky, Phys. Lett. B **688** (2009) 9.
[20] F. R. Urban, A. R. Zhitnitsky, Nucl. Phys. B **835** (2010) 135.
[21] N. Ohta, Phys. Lett. B **695** (2011) 41.
[22] S. Banerjee et al., Phys. Lett. B **611** (2005) 27.
[23] C. Pigozzo, M. A. Dantas, S. Carneiro, J. S. Alcaniz, JCAP **08** (2011) 022.
[24] L. M. Reyes, C. Moreno, J. E. Madriz-Aguilar, Eur. Phys. J Plus **127** (2012) 142.
[25] E. Komatsu, K. M. Smith et al., Astrophys. J. Suppl. **192** (2011) 18.
[26] H. A. Borges, S. Carneiro, Gen. Relativ. Gavit. **37** (2005) 1385.

Journal of Physics: Conference Series **545** (2014) 012011 doi:10.1088/1742-6596/545/1/012011

Non-Commutativity and the Theta Term

O Obregón[1] and R Santos-Silva[2]

Departamento de Física, Universidad de Guanajuato
C.P. 37150, León, Guanajuato, México.

E-mail: [1]octavio@fisica.ugto.mx

E-mail: [2]rsantos@fisica.ugto.mx

Abstract. In the presente paper we will study the non-commutativity in the space-time through the map developed by N. Seiberg and E. Witten, and explore the topological implications due to the introduction of non-commutativity. We will explore this topological properties using a non-commutative extension of the θ-term (second Chern class) on the 4-dimensional sphere.

1. Introduction

Non-commutativity in the space-time took N. Seiberg y E. Witten to construction of the well known Seiberg-Witten map [1]. Through the Moyal product they introduced non-commutativity on the space-time and fields; thereby the non-commutative gauge fields are an infinity serie on the non-commutative parameter $\theta_{\mu\nu}$ ($[x_\mu, x_\nu] = \theta_{\mu\nu}$ where x_μ are the coordinates on the space-time). But on the other hand assuming gauge invariance, it could be determined every order on the non-commutative field in terms of the usual gauge field (commutative field). This construction was realized for $U(1)$ gauge fields.

Such construction was extending by B. Jurco, L. Moller, S. Schraml, P. Schupp and J. Wess for non-abelian gauge fields coupled with matter [2], this proposal had been studied widely in the literature; in particular H. García-Compeán, O. Obregón, C. Ramírez and M. Sabido used this ideas to propose non-commutativity gravity, in a gauge field framework [4], also they extended the Euler characteristic and the Signature to their non-commutative counterparts [5].

Motivated by the idea of extending topological invariants on this non-commutative structure, the aim of this paper is to study the solution of instantons in Yang-Mills theories in the context of non-commutativity, i.e. we will focus on the study and definition of the non-commutative instanton number and which is the contribution or correction to the usual instanton number.

The structure of this paper is the following. We give two reviews: on the construction of the Seiberg-Witten map, continued by a review on the instanton solution on the S^4 equivalently $\mathbb{R}^4$ with topological charge equal 1. Then we give the definition of non-commutative θ-term and the computation up to second order, finally we present the conclusion and a discussion about the results.

Journal of Physics: Conference Series **545** (2014) 012011 doi:10.1088/1742-6596/545/1/012011

2. Non-Commutativity Via The Seiberg-Witten map

We are interested in non-commutativity in the space-time via the Seiberg-Witten map [1]. Such theory was extended by J. Wess [2, 3] for any gauge field coupled with matter. Below we present a brief description of the construction.

The central idea is to deform the algebraic structure of continuous spaces in particular the polynomials in N variables generated by powers of x^I where $I = 0, \dots, N$, which is a freely generated algebra $\mathbf{C}[\mathbf{x^1}, \dots, \mathbf{x^N}]$. Now consider the usual commutations relations between the coordinates

$$[x^\mu, x^\nu] = 0, \tag{1}$$

this algebraic structure will be deformed assuming

$$[\hat{x}^\mu, \hat{x}^\nu] = i\theta^{\mu\nu} \tag{2}$$

where $\theta^{\mu\nu} = -\theta^{\nu\mu} \in \mathbb{R}$ i.e. $\theta^{\mu\nu}$ is the non-commutative parameter and $\hat{x}$ are the non-commutative coordinates (as we can see this algebra is similar to the Heisenberg algebra in the phase space) for a formal description see [3].

Explicitly this modifies the way we multiply polynomials and in general functions over the non-conmutative variables in terms of the commutative variables through the $\star$ (Moyal) product defined by:

$$f \star g(x) = \mu(e^{\frac{i}{2}\Theta^{\rho\sigma}\partial_\rho \otimes \partial_\sigma} f \otimes g). \tag{3}$$

In this context the gauge transformations of a matter field $\Psi(x)$ is defined as

$$\delta_\Lambda \Psi(x) = i\Lambda \star \Psi(x), \tag{4}$$

$\Lambda(x)$ is the non-commutative gauge parameter which is Lie algebra valued i.e. $\Lambda(x) = \Lambda^a(x)T^a$. Now let us compute explicitly the following variation of the field Ψ

$$(\delta_{\Lambda_1}\delta_{\Lambda_2} - \delta_{\Lambda_1}\delta_{\Lambda_2})\Psi = (\Lambda_1 \star \Lambda_2 - \Lambda_2 \star \Lambda_1) \star \Psi = \frac{1}{2}\left([\Lambda_1^a, ^\star \Lambda_2^b]\{T^a, T^b\} + \{\Lambda_1^a, ^\star \Lambda_2^b\}[T^a, T^b]\right) \star \Psi. \tag{5}$$

It is worth mentioning that the fields are not Lie algebra valued because not only we have commutators but also we have anticommutators. So the algebra that close both operations (commutators and anticommutators) is known as universal enveloping algebra.

The covariant derivative is defined by

$$D_\mu^\star \Psi(x) = \partial_\mu \Psi(x) - i\hat{A}_\mu \star \Psi(x), \tag{6}$$

where $\hat{A}_\mu$ is the non-commutative gauge field which transform as

$$\delta_\Lambda \hat{A}_\mu = \partial_\mu \Lambda + i[\Lambda, ^\star \hat{A}_\mu]. \tag{7}$$

We can see that these terms have an infinity degrees of freedom, but N. Seiberg and E. Witten [1] showed that all the higher order terms depend only on the degrees at zeroth order (the commutative term) i.e. of the gauge parameter $\Lambda^{0\,a}T^a$ and the gauge field $A_\mu^{0\,a}T^a$. Let us assume that the gauge parameter Λ_α depends only on α and A_μ i.e. the gauge parameter and the gauge field respectively. With these assumptions bearing in mind, let us substitute in the expression (5)

$$\Lambda_\alpha \star \Lambda_\beta - \Lambda_\beta \star \Lambda_\alpha + i(\delta_\alpha \Lambda_\beta - \delta_\beta \Lambda_\alpha) = i\Lambda_{-i[\alpha,\beta]}. \quad (8)$$

This expression could be solved perturbatively assuming an expansion in the parameter θ as $\Lambda_\alpha = \Lambda^0_\alpha + \Lambda^1_\alpha + \cdots$, where $\Lambda^0_\alpha = \alpha = \alpha^a T^a$.

For example up to first order in θ, we find the following expression:

$$\delta_\alpha \Lambda^1_\beta - \delta_\beta \Lambda^1_\alpha - i[\alpha, \Lambda^1_\beta] - i[\Lambda^1_\alpha, \beta] - \Lambda^1_{-i[\alpha,\beta]} = -\frac{1}{2}\theta^{\mu\nu}\{\partial_\mu \alpha, \partial_\nu \beta\}. \quad (9)$$

whose solution is $\Lambda_\alpha = \alpha - \frac{1}{4}\theta^{\mu\nu}\{A_\mu, \partial_\nu \alpha\}$. With this expression we can compute in a similar way the expression for matter fields assuming the transformation at zeroth order $\delta_\alpha \Psi^0 = i\alpha\Psi^0$.

For the gauge field we expand again in orders of θ as $\hat{A} = A^0 + A^1 + A^2 + \cdots$ and substituting in (7), up to first order in θ, we obtain

$$A^1_\mu = \frac{1}{4}\theta^{\rho\sigma}(\{F_{\rho\mu}, A_\sigma\} - \{A_\rho, \partial_\sigma A_\mu\}), \quad (10)$$

finally the field strength tensor is given by $F^\star_{\mu\nu} = i[D^\star_\mu,^\star d^\star_\nu]$, whose solution up to first order is

$$F^\star_{\mu\nu} = F_{\mu\nu} + \frac{1}{4}\theta^{\rho\sigma}(2\{F_{\rho\mu}, F_{\sigma\nu}\} + \{D_\rho F_{\mu\nu}, A_\sigma\} - \{A_\rho, \partial_\sigma F_{\mu\nu}\}). \quad (11)$$

Where we can identify $F^0_{\mu\nu} = F_{\mu\nu}$ and $F^1_{\mu\nu} = -\frac{1}{4}\theta^{\rho\sigma}(2\{F_{\rho\mu}, F_{\sigma\nu}\} + \{D_\rho F_{\mu\nu}, A_\sigma\} - \{A_\rho, \partial_\sigma F_{\mu\nu}\})$.

3. Instantons

In this section we will review briefly the solution of instantons in S^4 (4-sphere) or equivalently in $\mathbb{R}^4$ in Yang-Mills theory (since S^4 is the one point compactification) we will follow closely the references [6, 7].

An instanton is a solution to the classical equations of motion of finite action namely, if A is an instanton solution which satisfies the boundary conditions $F_{\mu\nu} \to 0$ as $|x| \to \infty$ such that $S = \int_{S^4} F \wedge *F$ remains finitely, where $*$ is the hodge dual[1].

This implies from the mathematical point of view that the potential A is a self-dual or anti-self-dual connection (i.e. satisfies the instantons equations $*F = \pm F$) in a principal bundle over a 4-manifold.

For our case of interest the S^4 we will compute explicitly the potential that satisfies the mentioned boundary conditions and also under the anzats that the potential is purely gauge we assume that the potential has the following form

$$A_\mu(x) = f(x^2)g^{-1}(x)\partial_\mu g(x) \quad (12)$$

where f satisfies

[1] In physics $*F$ sometimes is called the dual of F and usually is denoted by $\tilde{F}$.

Journal of Physics: Conference Series **545** (2014) 012011 doi:10.1088/1742-6596/545/1/012011

$$f(x^2) \to 1 \qquad x^2 \to \infty, \tag{13}$$

we consider the following choice $g(x) = g_\alpha(x^2)\tau^\alpha$, where τ^α's are defined as

$$\tau^0 = I, \qquad \tau_j = i\sigma_j \quad \Rightarrow \quad \tau^{0\dagger} = \tau^0, \qquad \tau_j^\dagger = -i\sigma_j \tag{14}$$

with $j = 1, 2, 3$ and σ_j's are the Pauli matrices and I is de identity matrix 2×2. Since $g(x)$ is a $SU(2)$-valued function such that

$$g_\alpha \tau^{\alpha\dagger} g_\beta \tau^\beta = I. \tag{15}$$

Now consider $g_\alpha(x^2) = \partial_\alpha h(x^2) = 2x_\alpha h'(x^2)$, substituting in the previous expression and under the anzats we find

$$h(x^2) = (x^2)^{1/2} + c \quad \Rightarrow \quad g(x) = \frac{x_\alpha \tau^\alpha}{(x^2)^{1/2}}, \tag{16}$$

finally the potential has the following expression

$$A_\mu = \frac{f(x^2)}{x^2} x_\alpha \tau^{\alpha\dagger} \tau^\mu. \tag{17}$$

Substituting the potential into the field strength we find the condition for self-duality;

$$\partial_\mu \left(\frac{f(x^2)}{x^2} \right) + 2x_\mu \left(\frac{f(x^2)}{x^2} \right)^2 = 0, \tag{18}$$

solving the equation we find that

$$f(x^2) = \frac{x^2}{x^2 + \lambda^2} \qquad \text{where } \lambda \in \mathbf{R}, \tag{19}$$

finally substituting in (17) we get the desired solution.

$$A_\mu = \frac{x^2}{x^2 + \lambda^2} x_\alpha \tau^{\alpha\dagger} \tau^\mu. \tag{20}$$

3.1. Instanton Number

The instanton number or the first Chern class, also know in quantum field theory as the θ-term which is a topological term, is given by the following expression

$$k = -\frac{1}{8\pi^2} tr \int_{\mathbb{R}^4} F \wedge F, \tag{21}$$

so from the self-dual connection (20) we compute the field strength given explicitly by

$$F_{\mu\nu} = \frac{\lambda^2}{(x^2 + \lambda^2)^2} (\tau_\mu^\dagger \tau_\nu - \tau_\nu^\dagger \tau_\mu) \tag{22}$$

substituting in the previous expression we obtain $k = 1$, where we made use of the following relations

$$[\tau_\mu^\dagger, \tau_\nu] = 2\epsilon_{0\mu\nu\rho}\tau_\rho, \; \{\tau_\mu^\dagger, \tau_\nu\} = 2\delta_{\mu\nu} I + 2(\delta_{0\mu}\tau_\nu - \delta_{0\nu}\tau_\mu). \tag{23}$$

The instanton number also is known as topological charge, since we can add to the Yang-Mills action and does not contributes to the dynamics, but it carries topological information about the bundle, but at quantum level it is relevant (the path integral has their maximum contribution located at the instanton space).

4. Non-Commutative Instanton Number

The aim of this section is explore the modification to the topology when we introduce non-commutativity (through the Seiberg-Witten map), in particular we are interested on the information provided by the instanton number. Since the non-commutative extension gives new terms, our goal is to check if this information is topological, new, relevant or any of previously mentioned.

The non-conmutative θ-term is defined in similar way as the Yang-Mills action, redefining the action via the Moyal product

$$\int_M F \wedge \star F = \int_M \epsilon^{\mu\nu\rho\theta} \hat{F}_{\mu\nu} \hat{F}_{\rho\theta}, \tag{24}$$

where $\hat{F}_{\mu\nu} = F^0_{\mu\nu} + F^1_{\mu\nu} + F^2_{\mu\nu} + \cdots$ is the non-commutative field strength, where every term in the expansion is determined by the zeroth order, which are determined as in the previous section through the Seiberg-Witten map. Explicitly up to second order the A^i's are given (for arbitrary order expansion check [8])

$$A^0_\mu = A_\mu, \tag{25}$$

$$A^1_\mu = -\frac{1}{4}\theta^{\kappa\lambda}\{A_\kappa, \partial_\lambda A_\mu + F_{\lambda\mu}\}, \tag{26}$$

$$A^2_\mu = -\frac{1}{8}\theta^{\kappa\lambda}\left(\{A^1_\kappa, \partial_\lambda A_\mu + F_{\lambda\mu}\} + \{A_\kappa, \partial_\lambda A^1_\mu + F^1_{\lambda\mu}\}\right)$$

$$-\frac{i}{16}\theta^{\kappa\lambda}\theta^{\rho\nu}[\partial_\rho A_\kappa, \partial_\nu(\partial_\lambda A_\mu + F_{\lambda\mu})], \tag{27}$$

and the F^i's are

$$F^0_{\mu\nu} = F_{\mu\nu}, \tag{28}$$

$$F^1_{\mu\nu} = -\frac{1}{4}\theta^{\kappa\lambda}(\{A_\kappa, \partial_\lambda F_{\mu\nu} + D_\lambda F_{\mu\nu}\} - 2\{F_{\mu\kappa}, F_{\nu\lambda}\}), \tag{29}$$

$$F^2_{\mu\nu} = -\frac{1}{8}\theta^{\kappa\lambda}(\{A_\kappa, \partial_\lambda F^1_{\mu\nu} + D_\lambda F^1_{\mu\nu} - i[A_\lambda, F_{\mu\nu}] + \frac{1}{2}\theta^{\eta\delta}\{\partial_\eta A_\lambda, \partial_\delta F_{\mu\nu}\}\}$$

$$+\{A^1_\kappa, \partial_\lambda F_{\mu\nu} + D_\lambda F_{\mu\nu}\} - 2\{F_{\mu\kappa}, F^1_{\nu\lambda}\} - 2\{F^1_{\mu\kappa}, F_{\nu\lambda}\})$$

$$-\frac{i}{16}\theta^{\kappa\lambda}\theta^{\eta\delta}([\partial_\eta A_\kappa, \partial_\delta(\partial_\lambda F_{\mu\nu} + D_\lambda F_{\mu\nu})] - 2[\partial_\eta F_{\mu\kappa}, \partial_\delta F_{\nu\lambda}]). \tag{30}$$

the covariant derivative is $D_\lambda \Phi = \partial_\lambda \Phi - i[A_\lambda, \Phi]$ where Φ is a Lie algebra valued field.

Now consider the expansion of the θ-term (24) up to second order,

$$S = \int tr\epsilon^{\mu\nu\rho\theta}\left(F^0_{\mu\nu}F^0_{\rho\theta} + F^0_{\mu\nu}F^1_{\rho\theta} + F^1_{\mu\nu}F^0_{\rho\theta} + F^0_{\mu\nu}F^2_{\rho\theta} + F^1_{\mu\nu}F^1_{\rho\theta} + F^2_{\mu\nu}F^0_{\rho\theta}\right) d^4x, \tag{31}$$

using the expressions for the potential (self-dual connection) (20) and the field strength (22) with the aid of the expression for the different orders for the potential and field strength, we can compute the second order action for a $k = 1$ instanton over S^4.

For this case of $SU(2)$, we can check from the non-commutative action and the expression for the different orders; in general only the even orders contribute, since the even orders fields

Journal of Physics: Conference Series **545** (2014) 012011 doi:10.1088/1742-6596/545/1/012011

strength are always Lie algebra valued i.e. any even term will be a product of two Pauli matrices $\sigma_a\sigma_b = i\epsilon_{abc}\sigma_c + \delta_{ab}I$, hence when we take the trace this terms in general are not zero.

For the odd terms we deduce that the field strength of one of the products is proportional to an anticommutator between Pauli matrices i.e. in general always will be proportional to the identity ($\{\sigma_a, \sigma_b\} = 2\delta_{ab}I$), and when we multiply by the other Pauli matrices, this always be proportional to another Pauli matrices; hence the product in general will be always a Pauli matrix, therefore when we trace an odd term this term will be zero.

Thus the action (31) reduces to:

$$S = \int_{S^4} tr\epsilon^{\mu\nu\rho\theta} \left(F^0_{\mu\nu}F^0_{\rho\theta} + F^0_{\mu\nu}F^2_{\rho\theta} + F^1_{\mu\nu}F^1_{\rho\theta} + F^2_{\mu\nu}F^0_{\rho\theta} \right) d^4x, \tag{32}$$

where the first term is the usual commutative term $\int_{S^4} tr\epsilon^{\mu\nu\rho\theta}F^0_{\mu\nu}F^0_{\rho\theta} = 1$.

To compute the last three terms we use the equations (25)-(30) and substituting the expression for the potential and the field strength at zeroth order (20) and (22) for the solution of instanton ($k = 1$) respectively. It turns out that these terms (second order terms) are zero. Therefore there is no contribution up to second order, these means that the commutative θ-term and the non-commutative θ-term at least up to second order give us the same information.

5. Conclusions

We computed up to second order on S^4 (equivalently for $\mathbb{R}^4$) the non-commutative and the commutative θ-term and we observe that both have the same information; in this special case considered (up to second order), the instanton number was 1 in both cases.

It is worth mentioning that if we integrate the $k = 1$ solution for higher orders in θ, regardless the terms of the algebra it is clearly that the value of the integral over S^4 depends on λ (the instanton size), this suggest that these terms must be zero because they are not topological (since it depends on λ). Thus we expect that for orders larger than two each term must be zero, in the particular case S^4 there is no new topological information given through the non-commutative θ-term. This is in agreement with some results and discussion presented in [9].

It will be interesting to show formally in general for S^4 that the contribution of the non-commutativity θ-term is the same as the usual. By the other hand it could be more interesting compute the instanton number in a torus or in a different topological configuration.

References

[1] Seiberg N. and Witten E. *String Theory and Nonconmutative Geometry,* JHEP 9909:032 (1999).

[2] Jurco B., Moller L., Schraml S., Schupp P. and Wess J., *Construction of non-Abelian gauge theories on noncommutative spaces* Eur. Phys. J. C 21 (2001).

[3] Aschieri P., Dimitrijevic M., Kulish P., Lizzi F., Wess J., *Noncommutative Spacetimes: Symmetries in Noncommutative Geometry and Field Theory,* Lect. Notes Phys. 774 (Springer, Belin Heidelberg 2009).

[4] Garcia-Compean H., Obregon O., Ramirez C., Sabido M.,*Noncommutative Selfdual Gravity,* Phys.Rev. D68 (2003).

[5] Garcia-Compean H., Obregon O., Ramirez C., Sabido M., *Noncommutative Topological Theories of Gravity,* Phys.Rev. D68 (2003).

[6] Rajaraman R., *Solitons and Instantos,* North-Holland (1989).

[7] Nash C. and Sen S., *Topology and Geometry for Physicist,* Academic Press (1987).

[8] Ülker K. and Yapiskan B., *Seiberg-Witten maps to all orders,* Phys. Rev. D77 (2008).

[9] Madea Y., and Sako A., *Noncommutativity Deformation of Instantons* J. Geom. Phys. Vol. 58(12) (2008).

Vanishing condition for the heat flux of a relativistic fluid in a moving frame

Martín Romero-Muñoz
Physics Department, Universidad Autónoma Metropolitana-Iztapalapa, San Rafael Atlixco 186, México D. F. 09340, Mexico
E-mail: mrm@xanum.uam.mx

Leonardo Dagdug
Physics Department, Universidad Autónoma Metropolitana-Iztapalapa, San Rafael Atlixco 186, México D. F. 09340, Mexico
E-mail: dll@xanum.uam.mx

Guillermo Chacón-Acosta
Applied Mathematics and Systems Departament, Universidad Autónoma Metropolitana-Cuajimalpa, Av. Vasco de Quiroga 4871, México D. F. 05348, Mexico
E-mail: gchacon@correo.cua.uam.mx

Abstract. It has been asked if is appropriate to introduce the heat flow in the energy-momentum tensor, due to the non-mechanical nature of heat [1]. Although this could be answered by both kinetic and symmetry arguments, we address the problem by checking the validity of the second law of thermodynamics in a fluid that is boosted by a Lorentz transformation to a non comoving frame. In this contribution we found that this only can happen under certain conditions. Indeed, we found that there are a family of reference frames that satisfies these conditions, where Landau-Lifshitz frame is one of those. Additionally we relate such conditions with the null energy condition and the entropy production.

1. Introduction

The description of relativistic fluids is need to describe astrophysical systems such as neutron stars or primordial gases in cosmological models and in experiments such as collisions between heavy nuclei at high speeds at RHIC or LHC at CERN and recently to model systems such as graphene [2].

Heat flow and its constitutive equations in relativistic systems have been the subject of various interpretations and discussions in Eckart's formalism, where the constitutive equation of heat flow contains an acceleration term that seems to be the major source of instability in the fluctuations, which implies for instance, that water at room temperature becomes unstable and the nonexistence of the Rayleigh-Brillouin spectrum [3, 4].

Recently, constitutive equations that attempt to solve this problem have been proposed [4, 5, 6]. One alternative, comes from the relativistic kinetic theory [7] where a pressure gradient appears naturally instead of the acceleration term, and also turns theory stable [6, 4]. In the mid 80's the relativistic extension of the Meixner Prigogine formalism was made [8]. In this theory the heat flux is not included in the energy-momentum tensor [9], it appears as an independent 4-vector and as a consequence such unstabilites does not appear. It was argued that the only reason to include the heat flux in the energy momentum tensor is to fulfill the energy conservation [1]. It was claimed because the heat is a non mechanical kind of energy and also has properties of momentum and inertia, can not be included into energy-momentum tensor.

Given such objections the question whether to introduce the heat flux in the energy momentum tensor is in order. For its transformation properties [10] and also from the kinetic theoretical point of view the heat flux must be part of the energy tensor [7]. However we address this question from a rather simple point view.

Consider two inertial frames connected by a Lorentz transformation such that in the comoving frame the fluid has heat flow and in the moving one it does not. This might be interpreted that through a transformation between reference systems one could convert non-mechanical energy into purely mechanical. Apparently, this fact contradicts the second law of thermodynamics [11]. On the other hand, it can be thought that the frame with no heat flux is just the Landau-Lifshitz frame, the so-called frame *that moves with the heat*, however, as we shall see this frame has very strong conditions that are not essential.

In this work we found that, in order to cancel the heat flux with a Lorentz boost, some conditions must be satisfied by the fluid. We obtain those conditions and relate them with the null energy condition for the positivity of the energy density [12]. We also compute the entropy production and notice that is in agreement with the second law. We found that for a fluid that fulfill the conditions there are a plenty of reference frames in which the heat flow vanishes, one such is used in the Landau-Lifshitz approach.

2. Convective and non-convective energy flows

From kinetic arguments there is no objection to introduce the heat flux in the energy-momentum tensor [4, 7]. However there is a subtlety to consider when comparing with the non-relativistic case.

In case the non-relativistic energy flow can be written as [13]:

$$\frac{1}{2}\int v_i' v'^2 f dv' = u_i\left(\frac{1}{2}nmu^2+e\right)+u_i p_{ij}+\frac{1}{2}\int c_i' c'^2 f dv', \tag{1}$$

where $f=f(v')$ is the distribution function, v' is the molecular velocity, u is the hydrodynamic velocity molecules, c' the chaotic velocity (which is the molecular velocity in the comoving frame), nm mass density, e the energy and p the pressure.

There are three terms in (1): the first is the energy flow due to macroscopic convection and the second should be interpreted macroscopically due to work done by the stresses per unit time. The third term represents another kind of energy flow, which is usually called the heat flux vector, and is denoted by q:

$$q_i=\frac{1}{2}\int c_i' c'^2 f dv', \tag{2}$$

where it plays the same role in the macroscopic equations [13]. However, the name heat flow is somewhat misleading, because there are situations when $q\neq 0$ and the temperature is practically constant everywhere, in this case one has to talk of heat flux at constant temperature. Hence, the name non-convective energy flow would be more appropriate for q but is not used.

In the special relativistic case, we notice that by taking the Lorentz transform of the energy-momentum tensor to a moving frame, we obtain both a convective and a non-convective flux.

Journal of Physics: Conference Series **545** (2014) 012012 doi:10.1088/1742-6596/545/1/012012

Thus, it seems that we are converting non-mechanical to mechanical kind of energy. If a Lorentz transformation become all non-mechanical contribution, purely mechanical, we could be facing a system that does not satisfy the contents of the second law of thermodynamics [11].

In next section we obtain the conditions to be met by a fluid for make this possible, and verify if second law is violated.

3. Vanishing heat flux in a moving frame

From the point of view of kinetic theory, the energy momentum tensor is identified as the second statistical moment of the distribution function [7],

$$T^{\mu\nu} = \int p^\mu p^\nu f \frac{d^3p}{p^0}, \tag{3}$$

where p^0 is the energy, p^μ the 4-momentum and f the relativistic distribution function. In terms of its components is written as follows

$$T^{\mu\nu} = \begin{pmatrix} ne & \vec{q}^T \\ \vec{q} & \mathbb{P} \end{pmatrix}, \tag{4}$$

with $\mathbb{P}$ the viscosity stress tensor, $\vec{q}$ the heat flux vector, and ne the energy density that transforms with two Lorentz factors, one for the energy per particle and another by the density [10]. For a perfect fluid in its matrix form is represented as,

$$T^{\epsilon\sigma} = \begin{pmatrix} ne & 0 & 0 & 0 \\ 0 & p_x & 0 & 0 \\ 0 & 0 & p_y & 0 \\ 0 & 0 & 0 & p_z \end{pmatrix},$$

where it can be considered p_x, p_y, p_z as anisotropic pressures.

The matrix form of the Lorentz transformation in the direction x is,

$$\Lambda^\alpha_\epsilon = \begin{pmatrix} \gamma & -\gamma\beta & 0 & 0 \\ -\gamma\beta & \gamma & 0 & 0 \\ 0 & 0 & 1 & 0 \\ 0 & 0 & 0 & 1 \end{pmatrix},$$

where the Lorentz factor is given by $\gamma = \frac{1}{\sqrt{1-\beta^2}}$, and $\beta = \frac{v}{c}$ and c is the velocity of light. The Lorentz transformed quantities of the energy momentum tensor can be calculated as follows

$$\overline{T}^{\alpha\beta} = \Lambda^\alpha_\epsilon \Lambda^\beta_\sigma T^{\epsilon\sigma}. \tag{5}$$

The heat flux is given by

$$\bar{q}_x = -\beta\gamma^2(en + p_x). \tag{6}$$

As the reader can see there is a convective flow induced by the Lorentz transformation.

Next we consider a fluid with a heat flux in x direction as seen in its comoving frame. Then we compute a Lorentz transform in x direction where the heat flux turns to be,

$$\bar{q}_x = [q_x\beta^2 - (ne + p_x)\beta + q_x]\gamma^2. \tag{7}$$

Comparing (6) and (7) we notice that the total heat flux in this frame can be split into two parts, a convective and non-convective flux, as follows,

$$\bar{q}_x = \bar{q}_C + \bar{q}_{NC} \tag{8}$$

where

$$\bar{q}_C = -\beta\gamma^2(ne + p_x)$$
$$\bar{q}_{NC} = q_x\gamma^2(1+\beta^2).$$

Now, we set to zero the heat flux in the transformed tensor, which would mean that through a transformation between the reference systems we could convert convective into non-convective flux. With this end, we seek for the conditions that the thermodynamical quantities must satisfy in order to guarantee the existence of β i.e. such that $0 < \beta < 1$, and therefore the corresponding Lorentz transformation.

It is therefore necessary solve the following equation,

$$\alpha\beta^2 - \beta + \alpha = 0, \tag{9}$$

that comes form Eq. (7), where $\alpha = q_x/(ne + p_x)$. The solution to this last equation is easy to find and can be seen in Figure 1, namely,

$$\beta(\alpha) = \frac{1 \pm \sqrt{1-4\alpha^2}}{2\alpha}. \tag{10}$$

For $0 < \beta \leq 1$, the function α should be $\alpha \leq 0.5$. In particular when $\alpha = 0.5 \Rightarrow \beta = 1$ and is well known that photons do not define a frame of reference. So, in order for the heat flux vanish in this frame, the fluid in the comoving frame, should satisfy the following condition

$$q_x < \frac{1}{2}(ne + p_x). \tag{11}$$

As it is well known to make contact with the Landau Lifshitz scheme form the Eckart's frame, is possible to define a Lorentz frame only if $(q_x/nh_E)^2 \approx 0$, this condition is contained in Eq. (11) when $(q_x/nh_E)^2 < 1/4$, where $h_E = e + p_x/n$ [7]. Is important to notice that this implies that there is more than one reference frame where $\bar{q} = 0$.

This condition is simple and seems that can be obtained from another restriction. The null energy condition states that the energy density of any matter distribution measured by any null observer must be non-negative [12], and this happens when,

$$T^{\mu\nu}k_\mu k_\nu \geq 0, \tag{12}$$

where k_μ, k_ν are null vectors. If we choose the vector $k^\nu = (c, c, 0, 0)$, and the energy-momentum tensor with heat flux, then with a little algebra we obtain that,

$$\frac{1}{2} \geq \frac{q_x}{ne + p_x}, \tag{13}$$

which is the same condition as (11).

In order to verify the validity of the second law on a fluid under condition (11), let us calculate the entropy production in the Eckart frame [7]. Because we are considering a non viscous heat conducting fluid, the entropy production is reduced to

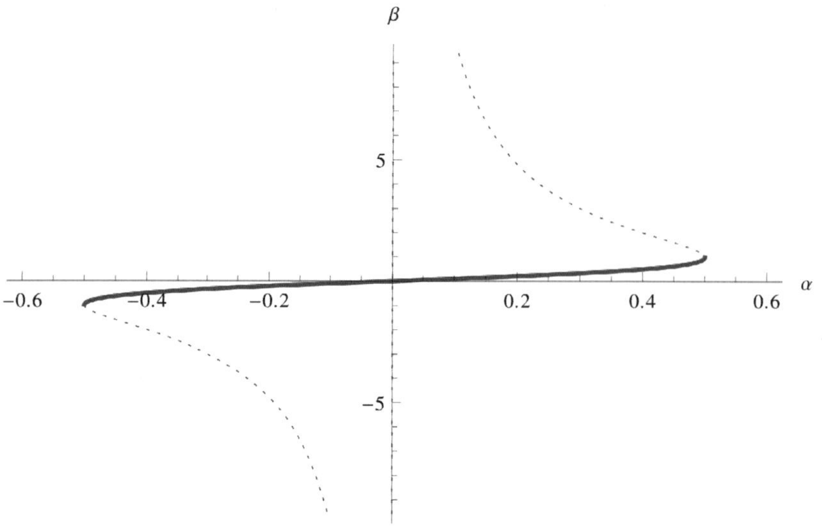

Figure 1. Figure shows that in order to has $0 < \beta < 1$, it is necessary that $0 < \alpha < \frac{1}{2}$, where $\alpha = \frac{q_x}{ne + p_x}$.

$$\varsigma = -\frac{1}{T^2}\left(q^\alpha \nabla_\alpha T - \frac{T}{c^2} q^\alpha D U_\alpha\right), \tag{14}$$

where T is the temperature, q^α the heat flux, c the speed of light, U_α the four velocity and D the convective time derivative. We also know that $q^\alpha = \Delta^\alpha_\gamma U_\beta T^{\beta\gamma}$, $\nabla_\alpha = \Delta^\beta_\alpha \partial_\beta$ and $D = U^\alpha \partial_\alpha$, where Δ^α_γ is the projector in the conmoving frame [7].

Assuming that the heat flux has only x component the entropy production takes the form,

$$\varsigma = -\frac{q_x c}{T^2}\left[\partial_x T - \frac{T}{c^2} D U_x\right]. \tag{15}$$

The term $-TDU_x/c^2$ in (15), has no correspondence in the non-relativistic case and represents the isothermal heat flux when the fluid is accelerated. This term acts in a direction opposite to the acceleration and is said to be due to the inertia of energy [15]. It is precisely this term which has caused controversy in the validity of Eckart's constitutive equation and seems to cause that generic instabilities [3].

We may express (15) in another way by eliminating the time derivative of the 4-velocity in the so-called Eulerian regime [7, 3],

$$\frac{nh_E}{c^2} D U^\alpha = \nabla^\alpha p, \tag{16}$$

where $h_E = e + p/n$. Then (15) takes the form,

$$\varsigma = -\frac{q_x c}{T^2}\left[\partial_x T - \frac{T}{nh_E}\partial_x p\right]. \tag{17}$$

Using condition (11) in the production of entropy (17), we have

$$\varsigma > -\frac{cnh_E}{2T^2}\partial_x T + \frac{1}{2T}\partial_x p, \tag{18}$$

which is positive and satisfies the second law, whenever,

$$\partial_x p > cnh_E \frac{\partial_x T}{T}. \tag{19}$$

4. Discussion

We study the heat flow behavior contained in the energy-momentum tensor under Lorentz transformations finding that there is a total heat flow that contains a convective and nonconvective part. When the heat flux is in the same direction as the Lorentz transformation was found that satisfying the condition (11), i.e. $q_x < \frac{1}{2}(ne + p_x)$ (this means that through a transformation between reference frames) is possible to convert some kind of convective energy in non-convective one. This condition can also be found through the null energy condition. When the Lorentz transformation is performed in a direction orthogonal to the heat flow it has a convective flow in the direction transformation and therefore the components do not mix each other. With the condition imposed in (11) is possible to find most of the reference systems, besides than the Landau-Lifshitz one for that the heat flux in a moving frame vanish. Moreover when we introduce the condition (11) in the entropy production it gives non-violation to the second law of thermodynamics. This happens only when the the pressure gradient is greater than the temperature gradient in the $x-$direction.

5. Acknowledgements

We thank the partial support from CONACyT grant No. 267825 (MRM), CONACyT grant under the project No. 176452 (LD) and project PROMEP 47510283 (GCA).

References

[1] L.S. García Colín, and A. Sandoval-Villalbazo, J. Non-Equilibrium Thermodynamics **31**, 11 (2006).
[2] M. Mendoza, N.A.M. Araújo, S. Succi, and H.J. Herrmann, Sci. Rep. **2** (2012) 611. A. Pototsky, F. Marchesoni, F. V. Kusmartsev, P. Hänggi, and S.E.Savelev, Eur. Phys. J. B **85** 356 (2012).
[3] A. L. García-Perciante, L. S. García-Colín y A. Sandoval-Villalbazo, Phys. Rev. E **79**, 066310 (2009).
[4] A. Sandoval Villalbazo, A.L. García Perciante, L.S. Garcia Colin, Physica A **388**, (2009).
[5] W. A. Hiscock, L. Lindblom, Phys. Rev. D **31** , 725 (1985).
[6] A.L. García Perciante, L.S. García-Colín, A. Sandoval-Villalbazo, General Relativity and Gravitation **41**,(2008).
[7] C. Cercignani y G. M. Kremer, *The Relativity Boltzmann Equation Theory and Applications* , Birkhäuser Verlag, Basel, Boston, Berlin, (2002).
[8] S.R. de Groot and P. Mazur, *Non-Equilibrium Thermodynamics* (Dover publications, Toronto, Ontario, 1984).
[9] L.S. GarcíaColín, and A. Sandoval-Villalbazo, Physica A **234** 358-370, (1996).
[10] B. Schutz, *A first course in general relativity* , 2nd edition, Cambridge University Press, (1985).
[11] L. S. García-Colín Scherer, *Introducción a la Termodinm̃ica Clásica*, (El Colegio Nacional, México 1990).
[12] R. M. Wald, *General Relativity*, (University of Chicago Press, 1984).
[13] C. Cercignani *Theory and application of the Boltzman equation*, Instituto di Matematica, Politecnico di Milano, Milano, Italy (1975).
[14] W. Greiner, Ludwig Neise, H. Stöcker *Thermodynamics and Statistical Mechanics*, Springer, (1995).
[15] C. Eckart, Phys. Rev. **58**, 919 (1940) .

Journal of Physics: Conference Series **545** (2014) 012013 doi:10.1088/1742-6596/545/1/012013

On the Chern-Simons State in General Relativity and Modified Gravity Theories

Ricardo Rosas-Rodríguez

Instituto de Física y Matemáticas, Universidad Tecnológica de la Mixteca, 69000, Huajuapan de León, Oaxaca, México

E-mail: rrosas@mixteco.utm.mx

Abstract. The Chern-Simons state is one solution to quantum constraints of gravity in the context of general relativity (GR) theory if we use Ashtekar's variables and if one orders the constraints with the triads to the left. Six years ago Krasnov introduced a certain class of modified gravity theories by replacing the cosmological constant by a "cosmological function of the curvature". If this function is a constant we come back to GR. In this note we review how the Chern-Simons state is one solution to the constraints of GR and we state the problem to face if we wish a generalized Chern-Simons state for the modified Krasnov's theories.

1. Introduction

In 2008 K. Krasnov introduced a class of alternative gravity theories with the name of "non-metric gravity", which are known as modified gravity theories today [1] (see also [2]). The essence of this modifies theories is to replace the cosmological constant by a "cosmological function" in the Hamiltonian constraint of the Ashtekar's canonical formulation of general relativity (GR) [3] (see also [4]). The Gauss and diffeomorphism constraints are unchanged. The canonically conjugate variables are still a densitized triad[1] $\tilde{\sigma}^a_i$ and a (complexified) SU(2) connection A^i_a. $a, b, c, \ldots$ are spatial indices and $i, j, k, \ldots$ are "internal" indices. The modified (Gauss, diffeomorphism and Hamiltonian) constraints of gravity take the following form [5]

$$\mathcal{G}_i = \mathcal{D}_a \tilde{\sigma}^a_i \approx 0, \tag{1}$$

$$\mathcal{H}_b = \tilde{\sigma}^{ai} F^i_{ab} \approx 0, \tag{2}$$

$$\mathcal{H} = \epsilon^{ijk} \tilde{\sigma}^{ai} \tilde{\sigma}^{bj} F^k_{ab} + \phi(\Psi_{\rm tf}) \epsilon^{ijk} \epsilon_{abc} \tilde{\sigma}^{ai} \tilde{\sigma}^{bj} \tilde{\sigma}^{ck} \approx 0, \tag{3}$$

where $\mathcal{D}_a$ is the covariant derivative with respect to the connection A^i_a, ϵ^{ijk} and ϵ_{abc} are the completely antisymmetric internal and spatial tensors taking values ± 1, $F^i_{ab} = 2\partial_{[a} A^i_{b]} + \epsilon^{ijk} A^j_a A^k_b$ is the curvature of A^i_a and $\phi(\Psi_{\rm tf})$ is an arbitrary function of the trace-free part of the tensor

$$\Psi^{ij} := \frac{F^i_{ab} \epsilon^{jkl} \tilde{\sigma}^{ak} \tilde{\sigma}^{bl}}{\epsilon_{ijk} \epsilon_{abc} \tilde{\sigma}^{ai} \tilde{\sigma}^{bj} \tilde{\sigma}^{ck}}. \tag{4}$$

[1] We use the Ashtekar convention: the tilde above a symbol denotes a density of weight 1, under a symbol a density of weight −1.

Note that $\Psi^{ij} \approx \Psi^{(ij)}$ in view of the diffeomorphism constraint (2). GR is the simplest case $\phi =$ constant $= \Lambda$, in which the quantity $\Psi^{ij}_{\rm tf}$ defined by (4) is nothing but the (self-dual part) of the Weyl curvature tensor [6].

The spacetime covariant theory that leads to the modified constraints (1), (2) and (3) was proposed in [2] with particular arguments about why this theory should be closed under renormalization and its action is given by

$$S = \int B_i \wedge F^i(A) - \frac{1}{2}\left(\Lambda^{ij} - \frac{1}{3}\delta^{ij}\phi(\Lambda)\right) B^i \wedge B^j, \tag{5}$$

where B^i is an SU(2) Lie algebra valued two-form, A^i is the connection one-form, $F^i(A)$ is its curvature, and Λ^{ij} is the Lagrange multiplier field. When $\phi =$ constant $= \Lambda$ one obtains from (5) the Plebański action for GR with the cosmological constant [7]. The $3+1$ decomposition of the theory (5) is given in [5].

In [5] and [8] it is shown that the constraints (1), (2) and (3) form a first-class algebra, i.e., that the Poisson bracket of any two constraints vanishes on the constraint surface. Therefore these constraints generate gauge transformations of the modified theories and the count of the number of physical degrees of freedom is simple: we have nine kinematical configurational variables A^i_a, together with seven constraints, which give two physical degrees of freedom.

The constraint algebra for the 2+1-dimensional modified gravity theories is studied in [9].

The aim of this note is to present the problem to face if we wish a generalization of the Chern-Simons state to the modified Krasnov's theories.

In the next section we show how the Chern-Simons state is one solution to the constraints of GR. In Sect. 3 we show the problem to face if we wish a generalized Chern-Simons state for the modified Krasnov's theories. We finish this letter with some concluding remarks.

2. Chern-Simons State for GR

If we ignore the condition of square integrability and think of states as arbitrary functions $\psi[A]$, to quantize, we replace the classical 'position' and 'momentum' variables A^i_a and E^a_i by the operators

$$\hat{A}^i_a\psi[A] = A^i_a\psi[A], \qquad \hat{E}^a_i\psi[A] = \frac{\delta}{\delta A^i_a}\psi[A]$$

which have commutation relations analogous to the Poisson bracket relations. With these operators in hand we can then make the Hamiltonian, diffeomorphism and Gauss law constraints into operators. There are different operator ordering to choose from, but we choice the triads to the left:

$$\begin{aligned} \hat{\mathcal{G}}_i &= \hat{D}_a\hat{E}^a_i \\ \hat{\mathcal{H}}_b &= \hat{E}^{ai}\hat{F}^i_{ab} \\ \hat{\mathcal{H}} &= \epsilon^{ijk}\hat{E}^{ai}\hat{E}^{bj}\hat{F}^k_{ab} - \frac{\Lambda}{6}\epsilon^{ijk}\epsilon_{abc}\hat{E}^{ai}\hat{E}^{bj}\hat{E}^{ck} \end{aligned}$$

where Λ is the cosmological constant and

$$\hat{F}^i_{ab}\psi[A] = F^i_{ab}\psi[A]. \tag{6}$$

We define the physical state space $\mathcal{H}_{\rm phys}$ to be the space of functions $\psi[A]$ that satisfy the constraints in quantum form, i.e.,

$$\mathcal{H}_{\rm phys} = \{\psi : \hat{\mathcal{G}}_i\psi = \hat{\mathcal{H}}_b\psi = \hat{\mathcal{H}}\psi = 0\}. \tag{7}$$

Then, the problem is to find functions ψ in $\mathcal{H}_{\text{phys}}$.

The Chern-Simons theory [11] gives rise to a solution of all three constraint equations, if we choice $\Lambda \neq 0$.

We define the Chern-Simons state Ψ_{CS} to be the following function

$$\Psi_{CS}[A] = e^{-\frac{6}{\Lambda}S_{CS}[A]} \tag{8}$$

where

$$\begin{aligned} S_{CS}[A] &= \int_\Sigma \text{tr}(A \wedge dA + \frac{2}{3}A \wedge A \wedge A) \\ &= \int_\Sigma \epsilon^{abc}\text{tr}\left(A_a\partial_b A_c + \frac{2}{3}A_a A_b A_c\right) \end{aligned} \tag{9}$$

is the Chern-Simons action.

Ψ_{CS} lies in the physical state space for quantum gravity with cosmological constant:

$$\hat{\mathcal{G}}_i\Psi_{CS} = \hat{\mathcal{H}}_b\Psi_{CS} = \hat{\mathcal{H}}\Psi_{CS} = 0. \tag{10}$$

Since the Chern-Simons action is invariant under gauge transformations and diffeomorphisms, two of these three equations are obvious: $\hat{\mathcal{G}}_i\Psi_{CS} = \hat{\mathcal{H}}_b\Psi_{CS} = 0$. The fact that $\hat{\mathcal{H}}\Psi_{CS} = 0$ is not so obvious. We offer a direct argument to show this.

First, the functional derivative of S_{CS} is

$$\frac{\delta}{\delta A_c^k}S_{CS}[A] = -\frac{1}{2}\epsilon^{abc}F_{abk}.$$

Then,

$$\begin{aligned} \frac{\delta}{\delta A_c^k}\Psi_{CS}[A] &= \frac{\delta}{\delta A_c^k}e^{-\frac{6}{\Lambda}S_{CS}[A]} = \frac{3}{\Lambda}\epsilon^{abc}F_{abk}e^{-\frac{6}{\Lambda}S_{CS}[A]} \\ &= \frac{3}{\Lambda}\epsilon^{abc}F_{abk}\Psi_{CS}[A]. \end{aligned}$$

It follows that

$$\epsilon_{abc}\frac{\delta}{\delta A_c^k}\Psi_{CS}[A] = \frac{6}{\Lambda}F_{abk}\Psi_{CS}[A].$$

Finally

$$\hat{\mathcal{H}}\Psi_{CS} = \epsilon^{ijk}\frac{\delta}{\delta A_a^i}\frac{\delta}{\delta A_b^j}\left(F_{abk} - \frac{\Lambda}{6}\epsilon_{abc}\frac{\delta}{\delta A_c^k}\right)\Psi_{CS} = 0.$$

The work of Kodama [12] indicate that the Chern-Simons state is a quantized version of anti-deSitter space, a simple solution of the vacuum Einstein equations with $\Lambda \neq 0$.

3. Chern-Simons State for Modified Gravity Theories

For Krasnov's modified gravity theories we define the modified Chern-Simons state as

$$\Psi_{MCS}[A] = e^{-\frac{6}{\phi(\Psi^{ij})}S_{CS}[A]} \tag{11}$$

Journal of Physics: Conference Series **545** (2014) 012013 doi:10.1088/1742-6596/545/1/012013

where $S_{CS}[A]$ is the Chern-Simons action again. Note that Ψ_{MCS} is yet a gauge and Lorentz scalar, thus, we have yet $\hat{\mathcal{G}}_i\Psi_{MCS} = \hat{\mathcal{H}}_b\Psi_{MCS} = 0$, directly. The trouble again is to show that $\hat{\mathcal{H}}\Psi_{MCS} = 0$.

In this case, we have

$$\begin{aligned}
\frac{\delta}{\delta A^k_c}\Psi_{MCS} &= \frac{\delta}{\delta A^k_c} e^{-\frac{6}{\phi(\Psi^{ij})}S_{CS}} \\
&= -\left[\frac{6}{\phi(\Psi^{ij})}\frac{\delta}{\delta A^k_c}S_{CS} + S_{CS}\frac{\delta}{\delta A^k_c}\frac{6}{\phi(\Psi^{ij})}\right] e^{-\frac{6}{\phi(\Psi^{ij})}S_{CS}} \\
&= \left[\frac{6}{\phi(\Psi^{ij})}\left(\frac{1}{2}\epsilon^{abc}F_{abk}\right) + S_{CS}\frac{6}{\phi^2(\Psi^{ij})}\frac{\delta\phi}{\delta A^k_c}\right]\Psi_{MCS}.
\end{aligned} \tag{12}$$

Thus

$$\epsilon_{abc}\frac{\delta}{\delta A^k_c}\Psi_{MCS} = \frac{6}{\phi(\Psi^{ij})}F_{abk}\Psi_{MCS} + S_{CS}\frac{6\epsilon_{abc}}{\phi^2(\Psi^{ij})}\frac{\delta\phi}{\delta A^k_c}\Psi_{MCS},$$

and it is not obvious that $\hat{\mathcal{H}}\Psi_{MCS} = 0$.

4. Concluding Remarks

We can see that $\hat{\mathcal{H}}\Psi_{MCS} = 0$ does not follow directly for modified gravity theories as for GR.

We must to point out that perhaps a different modified Chern-Simons state is need to be defined in order to solve the scalar constraint, but at the present we have not idea of the form of this state.

In this letter we only wish to show the trouble that emerge if we define the modified Chern-Simons state as in (11).

Acknowledgments

I am grateful to Professor Riccardo Capovilla for drawing my attention onto this matter. I would like to thank the organizers of the X Workshop on Gravitation and Mathematical Physics of the DGFM of the SMF for their hospitality.

References

[1] K. Krasnov, *Class. Quantum Grav.* **25**, 025001 (2008).
[2] K. Krasnov, Renormalizable non-metric gravity?, *Preprint* hep-th/0611182.
[3] A. Ashtekar, *Phys. Rev. Lett.* **57**, 2244 (1986); *Phys. Rev.* D **36**, 1587 (1987).
[4] R. Rosas-Rodríguez, *Int. J. Mod. Phys.* A **23**, 895 (2008).
[5] K. Krasnov, *Phys. Rev. Lett.* **100**, 081102 (2008).
[6] R. Capovilla, J. Dell, T. Jacobson and L. Mason, *Class. Quantum Grav.* **8**, 41 (1990).
[7] J. Plebańsky, *J. Math. Phys.* **18**, 2511 (1977).
[8] R. Rosas-Rodríguez, *AIP Conf. Proc.* **1287**, 85 (2010).
[9] R. Rosas-Rodríguez, *AIP Conf. Proc.* **1473**, 248 (2012).
[10] R. Capovilla, M. Montesinos and M. Velázquez, *Class. Quantum Grav.* **27**, 145011 (2010).
[11] J. Baez and J. P. Muniainn, *Gauge Fields, Knots and Gravity*, World Scientific, Singapore (1994).
[12] H. Kodama, *Int. J. Mod. Phys.* D **1**, 439 (1992).

Journal of Physics: Conference Series **545** (2014) 012014 doi:10.1088/1742-6596/545/1/012014

Effective potentials in geodesic curves on surfaces

J.A. Santiago[a], **G. Chacón-Acosta**[b] **and O. González-Gaxiola**[c]

Applied Mathematics and Systems Deptartment, Universidad Autónoma Metropolitana – Cuajimalpa, Vasco de Quiroga 4871, Santa Fe, Cuajimalpa, C.P. 05348, México D. F., MEXICO

E-mail: [a]jsantiago@correo.cua.uam.mx, [b]gchacon@correo.cua.uam.mx, [c]ogonzalez@correo.cua.uam.mx

Abstract. In this work, the equations of geodesic curves on surfaces embedded in euclidean space are obtained. By introducing a vector Lagrange multiplier, we show that the geodesic curvature of the curves are zero and the normal curvature of them can be identified with the force transmitted to the surface. We then obtain the corresponding formulas in the case of axially symmetric surfaces, where a first integral of the geodesic equations can be interpreted as a particle moving in an effective potential (being zero the total energy), and the angular momenta is conserved. The methodology developed is illustrated with some examples: the catenoid and the pseudosphere.

1. Introduction

In geometry, a geodesic curve is defined as the curve of minimum length (and less curvature) joining two given points on a given surface. Geodesics are one of the families of curves often called special lines or curves. The term comes from the word geodesy, science to measure the size and shape of the Earth. The formal study of the geodesic starts the year 1691, when John Bernoulli proposed the brachistochrone problem [1]. In 1731 A.C. Clairaut published a treatise on curves of double curvature, *Recherches sur les courbes a double curvature*, which allowed his admission to the French Academy of Sciences [2]; subsequently, C.F. Gauss (1777-1855) emphasizes the geometric nature of the mathematical analysis, with mainly two contributions: the birth of complex analysis and the differential geometry analysis [2]. In 1866 Jean Gaston Darboux in his doctoral thesis began a prolific development on geometry, some results published in two articles in the journal *Annales de l'Ecole Normale Suprieure de Paris*, Darboux was the first to characterize a curve on a surface (given its curvature and torsion) by the differential equation, this was done around 1871. Particularly a surface of revolution is a surface which is generated from the rotation of a planar curve around an axis which lies in the same plane, and is a Liouville surface.

Although the analysis of geodesic curves appear in any book of geometry, most of the analysis is done intrinsically, especially as an application of variational calculus [3]. One of the main applications of geodesic analysis is given in physics where free particles follow geodesic trajectories [4]. Recently it has been shown that, in the interaction of crystalline order on curved surfaces, there is a non-local effect that forces normal lines to be on geodesic curves [5]. If one includes the bending energy, the equilibrium configurations of elastic polymers on surfaces as spheres can be found [6]. Indeed it is even possible to include both long range forces between the curve itself, as well as between the curve and the surface [7].

2. Geodesic curves confined on surfaces

Let $\mathbf{x} = (x^1, x^2, x^3)$, be a point $\in \mathbb{R}^3$, and $\mathbf{x} = \mathbf{Y}(s)$, a curve parametrized by s, restricted to be along the surface $\mathbf{x} = \mathbf{X}(\xi^a)$, parametrized by local coordinates $\xi^a, a = 1, 2$. To take into account this requirement, we follow [6] and introduce a vector Lagrange multiplier $\boldsymbol{\lambda}$, so that the length functional is given by

$$L(\mathbf{Y}, \xi^a) = \int ds + \int \boldsymbol{\lambda} \cdot [\mathbf{Y}(s) - \mathbf{X}(\xi^a)]\, ds, \tag{1}$$

where the dot denotes the internal product in $\mathbb{R}^3$. In order to find the Euler-Lagrange equations we have to consider deformations δL as a consequence of deformations of the variables $\delta\mathbf{Y}$ and $\delta\xi^a$. We obtain

$$\delta L(\mathbf{Y}, \xi^a) = \int \frac{d}{ds}(\delta\mathbf{Y}) \cdot \mathbf{T} ds + \int \boldsymbol{\lambda} \cdot \delta\mathbf{Y}, \tag{2}$$

where $\mathbf{T} = d\mathbf{Y}/ds = \dot{\mathbf{Y}}$ is the tangent vector field to the curve. By doing one integration by parts, we have that $\delta L = 0$, implies that,

$$\dot{\mathbf{T}} = \boldsymbol{\lambda}. \tag{3}$$

Variation respect to the variable ξ^a gives $\delta L = -\int \boldsymbol{\lambda} \cdot \mathbf{e}_a \delta\xi^a$, where $\mathbf{e}_a = \partial_a \mathbf{X}$, are the two tangent vectors to the surface. We thus see that $\delta L = 0$ implies that $\boldsymbol{\lambda} = -\lambda\mathbf{n}$, being $\mathbf{n}$ the unit normal to the surface. Projection of $\dot{\mathbf{T}}$ along the Darboux basis $\{\mathbf{T}, \mathbf{n}, \mathbf{l} = \mathbf{T} \times \mathbf{n}\}$, gives us

$$\begin{aligned} \dot{\mathbf{T}} &= -K_{ab} t^a t^b \mathbf{n} + \left(\dot{t}^c + \Gamma^c_{ab} t^a t^b\right) \mathbf{e}_c, \\ &= -\kappa_n \mathbf{n} + \kappa_g \mathbf{l}, \end{aligned} \tag{4}$$

where the normal curvature $\kappa_n = K_{ab} t^a t^b$ with $t^a = d\xi^a/ds$, and the geodesic curvature $\kappa_g = \left(\dot{t}^c + \Gamma^c_{ab} t^a t^b\right) l_c$ has been introduced. One conclude from equation (3) that $\kappa_g = 0$ for geodesic curves, as it should be. We also see that $\lambda = \kappa_n$, can be interpreted as the force transmitted to the surface. Some examples about this curves and forces are given below.

3. Surfaces of revolution

A surface of revolution can be described parametrically through

$$\mathbf{X}(\varphi, v) = (\rho(v)\cos\varphi, \rho(v)\sin\varphi, z(v)). \tag{5}$$

The infinitesimal element of distance along the surface can be written as $ds^2 = \rho^2 d\varphi^2 + (\rho'^2 + z'^2)dv^2$, where – ' – indicates derivative respect to v. The two tangent vectors $\mathbf{e}_a$ are given by

$$\begin{aligned} \mathbf{e}_\varphi &= (-\rho\sin\varphi, \rho\cos\varphi, 0), \\ \mathbf{e}_v &= (\rho'\cos\varphi, \rho'\sin\varphi, z'), \end{aligned}$$

whereas the normal unit vector to the surface can be written in the form,

$$\mathbf{n} = \frac{1}{\sqrt{\rho'^2 + z'^2}}(z'\cos\varphi, z'\sin\varphi, -\rho'). \tag{6}$$

The second fundamental form K_{ab} of the surface can be calculated through its definition in the Gauss equation, $\partial_a \mathbf{e}_b = -K_{ab}\mathbf{n}$. As a consequence of the cylindrical symmetry a diagonal matrix is found for the extrinsic curvature, being $K_{\varphi\varphi} = \frac{\rho z'}{\sqrt{\rho'^2+z'^2}}$ and $K_{vv} = \frac{\rho' z'' - \rho'' z'}{\sqrt{\rho'^2+z'^2}}$, its nonzero

Journal of Physics: Conference Series **545** (2014) 012014 doi:10.1088/1742-6596/545/1/012014

elements. Being also diagonal the matrix $K_a^{\ b}$, their components are the principal curvatures of the surface, $K^\varphi_\varphi = \frac{z'}{\rho\sqrt{\rho'^2+z'^2}}$ and $K^v_v = \frac{K_{vv}}{\rho'^2+z'^2}$. The trace of the second fundamental form $K = g^{ab}K_{ab}$ is given by

$$K = \frac{1}{\sqrt{\rho'^2+z'^2}}\left(\frac{z'}{\rho} + \frac{\rho' z'' - \rho'' z'}{\rho'^2+z'^2}\right). \tag{7}$$

The curvature scalar $\mathcal{R}$ can be obtained intrinsically using the induced metric g_{ab} through the Christoffel symbols. Nevertheless, we can calculate it, through the integrability condition given by the Codazzi identity identity $\mathcal{R} = K^2 - K^{ab}K_{ab}$, so we have

$$\mathcal{R} = \frac{2z'}{\rho}\frac{\rho' z'' - \rho'' z'}{(\rho'^2+z'^2)^2}. \tag{8}$$

In addition, because the symmetry under rotations around the z axis, $\ell = \mathbf{L}\cdot\mathbf{k}$ is conserved along the curve, where $\mathbf{L} = \mathbf{x}\times\dot{\mathbf{x}}$, is the angular momenta. We find

$$\ell = \rho^2\frac{d\varphi}{ds}. \tag{9}$$

With these elements, we can write the normal curvature along the trajectory of the particle as follows

$$\kappa_n = K_{\varphi\varphi}\left(\frac{d\varphi}{ds}\right)^2 + K_{vv}\left(\frac{dv}{ds}\right)^2. \tag{10}$$

Note also that, using the infinitesimal element of distance we can get a first integral of the geodesic equation

$$\frac{1}{2}\left(\frac{dv}{ds}\right)^2 + U_{eff} = 0, \tag{11}$$

where

$$U_{eff} = -\frac{1}{2(\rho'^2+z'^2)}\left(1 - \frac{\ell^2}{\rho^2}\right), \tag{12}$$

can be interpreted as an effective potential that depends on the coordinates and on the conserved angular momentum explicitly. Then we can write the normal curvature along geodesic curves as

$$\kappa_n = \frac{\ell^2}{\rho^3}\frac{z'}{\sqrt{\rho'^2+z'^2}} + \frac{\rho' z'' - \rho'' z'}{(\rho'^2+z'^2)^{3/2}}\left(1 - \frac{\ell^2}{\rho^2}\right), \tag{13}$$

thus, the force λ is completely determined.

3.1. Catenoid

In this case we have $\mathbf{X}(\varphi, v) = (\cosh v\cos\varphi, \cosh v\sin\varphi, v)$. Then we identify $\rho(v) = \cosh v$ and $z(v) = v$. We have $K_{\varphi\varphi} = 1$ and $K_{vv} = -1$, and the trace $K = 0$ (a minimal surface), its curvature $\mathcal{R} = -2\,\mathrm{sech}^4 v < 0$. The effective potential of the fictitious particle (of total energy zero) is given by

$$U_{eff}(v) = -\frac{1}{2}\mathrm{sech}^2 v\left(1 - \ell^2\mathrm{sech}^2 v\right). \tag{14}$$

Figure (1) shows the behavior of geodesic curves: For $0 < \ell < 1$, geodesics crosses parallels along the catenoid; for $\ell > 1$, the potential barrier does not allow the geodesic going from one side to another of the catenoid. The equator corresponds to the geodesic with $\ell = 1$, the geodesic with $\ell = 0$ is a meridian. The transmitted force to the surface in this case reads as follows

$$\lambda_\ell(v) = \mathrm{sech}^2 v\left(2\ell^2\mathrm{sech}^2 v - 1\right). \tag{15}$$

In Fig. (2), the force λ, transmitted to the catenoid is shown. For curves crossing the catenoid, such that $\ell \geq 1/\sqrt{2}$, there exist v_0 where the force transmitted to the surface is zero, $\lambda_\ell(v_0) = 0$.

Journal of Physics: Conference Series **545** (2014) 012014 doi:10.1088/1742-6596/545/1/012014

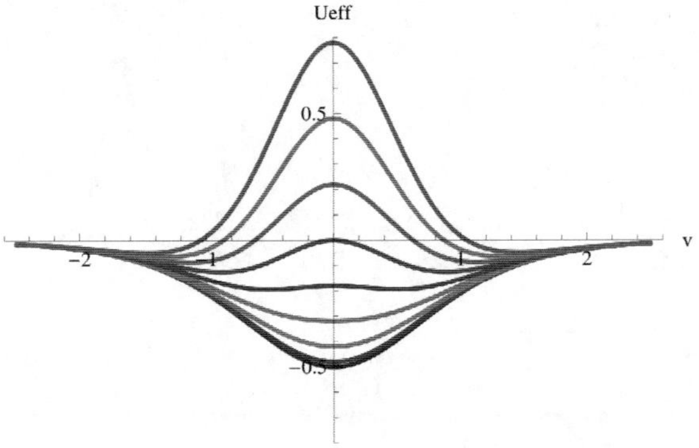

Figure 1. The effective potential (14) of the catenoid, for several values of the angular momenta ℓ: the bottom curve correspond to $\ell = 0$. After reaching $\ell = 1$, a barrier appears.

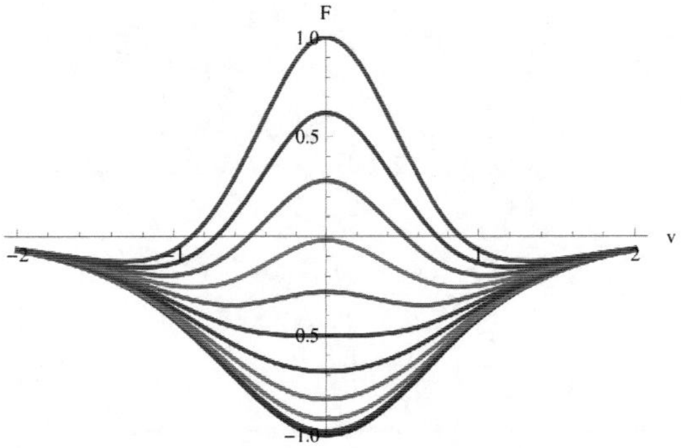

Figure 2. The transmitted force to the catenoid for several values of the angular momenta ℓ: the bottom curve correspond to $\ell = 0$, the upper one to $\ell = 1$.

3.2. Pseudosphere

A parametrization of the pseudosphere is given by $\mathbf{X}(\varphi, v) = (\text{sech}v \cos\varphi, \text{sech}v \sin\varphi, v - \tanh v)$. We identify then $\rho(v) = \text{sech}v$, and $z(v) = v - \tanh v$. The non-zero components of the second fundamental are given by $K_{\varphi\varphi} = \text{sech}v \tanh v$ and $K_{vv} = -K_{\varphi\varphi}$. The trace is given by

$$K = \frac{(\cosh 2v - 3)}{2\sinh v}. \tag{16}$$

As it is well known, the pseudosphere is a surface of negative curvature with $\mathcal{R} = -2$. The effective potential is now given by

$$U_{eff}(v) = -\frac{1}{2\tanh^2 v}\left(1 - \frac{\ell^2}{\text{sech}^2 v}\right). \tag{17}$$

Figure (3) shows the behavior of pseudosphere's geodesic curves: For $0 < \ell < 1$, geodesics remain on one side of the surface; for $\ell \geq 1$, the potential barrier does not allow the geodesic to exist. Meridians are geodesic curves ($\varphi = \text{const.}$) along that $\ell = 0$.

The force transmitted to the surface can be written in the form

$$\lambda_\ell(v) = \frac{\ell^2 \cosh^3 v - \text{sech}v}{\tanh v}. \tag{18}$$

4. Conclusions

By introducing a vector Lagrange multiplier into the length functional, we obtain the geodesic curves on surfaces embedded in euclidian space, through the obtention of the Euler-Lagrange equations. We found that the geodesic curvature vanishes for geodesic curves and the normal curvature is directly related with the transmitted force to the surface. We studied two axially symmetric surfaces, namely, the catenoid and the pseudosphere, where a first integral of the geodesic equation can be thought as a particle (with vanishing energy) moving in an effective potential U_{eff}. In such a cases the force or the normal curvature are completely determined and turn out to be a function of the conserved angular momenta.

Acknowledgments

JAS would like to thank Jemal Guven the many lessons on geometry.

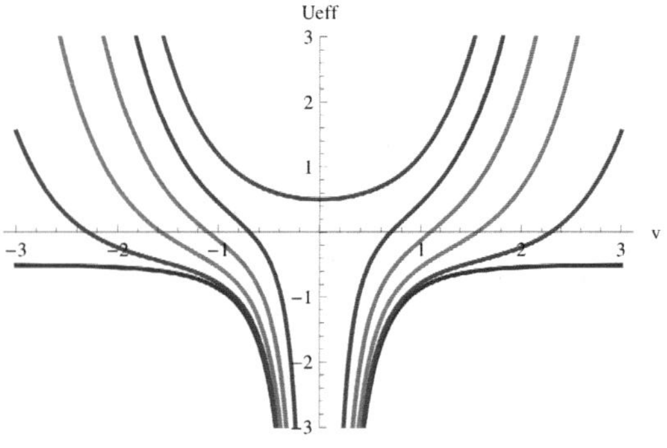

Figure 3. The effective potential (14) of the pseudosphere for several values of the angular momenta ℓ: the bottom curve corresponds to $\ell = 0$. The upper one to $\ell = 1$.

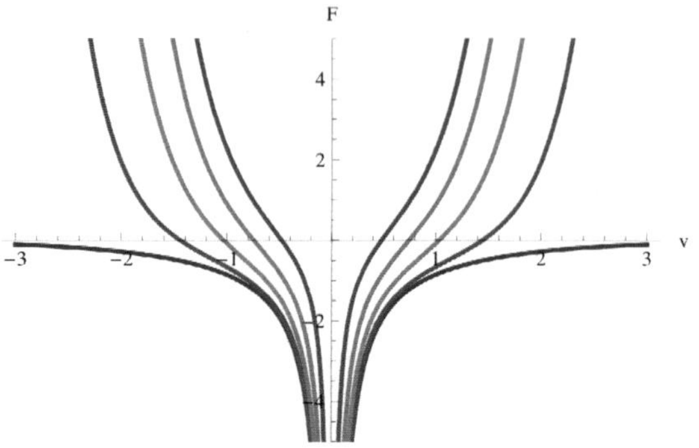

Figure 4. The transmitted force to the catenoid for several values of the angular momenta ℓ, the bottom curve corresponds to $\ell = 0$.

References

[1] Struik DJ 1965 *A Concise History of Mathematics*, (Dover Books on Mathematics)
[2] Collete JP 1985 *Historia de las matemáticas vol. II* (Siglo XXI de España Editores)
[3] do Carmo M 1976 *Differential Geometry of Curves and Surfaces* (Prentice-Hall)
[4] Hartle JB 2003 *Gravity: An introduction to Einstein's General Relativity* (Addison Wesley)
[5] Kamien RD, Nelson DR, Santangelo CD, Vitelli V 2009 *Phys. Rev.* E **80** 051703
[6] Guven J and Vazquez-Montejo P 2013 *Phys. Rev.* E **85** 026603
[7] Santiago JA, Chacón-Acosta G and González-Gaxiola O 2013 *IJMP* B **27** 1350043

Anisotropic cosmology in K-essence theory

J Socorro[1,2], Luis O Pimentel[2], Abraham Espinoza-García[1]

[1]Departamento de Física de la DCeI de la Universidad de Guanajuato-Campus León
C.P. 37150, Guanajuato, México

[2]Departamento de Física de la Universidad Autónoma Metropolitana
Apartado Postal 55-534, 09340, México, D.F.

E-mail: socorro@fisica.ugto.mx

E-mail: lopr@xanum.uam.mx

E-mail: abraham@fisica.ugto.mx

Abstract. We use one of the simplest forms of the K-essence theory and we apply it to the classical anisotropic Bianchi type I cosmological model, with a barotropic perfect fluid ($p = \gamma\rho$) modeling the usual matter content and include the particular form of potential $V(\phi) = \text{constant} = 2\Lambda$. The classical solutions for any $\gamma \neq 1$ and $\Lambda = 0$ are found in closed form, using a time transformation. We also present the solution when $\Lambda \neq 0$ including particular values in the barotropic parameter. We present the possible isotropization of the cosmological model Bianchi I using the ratio between the anisotropic parameters and the volume of the universe and show that this tend to a constant or to zero for different cases.

Keywords: K-essence theory; anisotropic model; classical solution;

1. Introduction

In recent times, some attempts to unify the description of dark matter, dark energy and inflation, by means of a scalar field with non standard kinetic term have been conducted [1, 2, 3, 4]. The K-essence theory is based on the idea of a dynamical attractor solution which causes it to act as a cosmological constant only at the onset of matter domination. Consequently, K-essence overtakes the matter density and induces cosmic acceleration at about the present epoch. Usually K-essence models are restricted to the lagrangian density of the form [2, 5, 6, 7]

$$S = \int d^4x \sqrt{-g}\, [f(\phi)\, \mathcal{G}(X) - V(\phi)]\,, \tag{1}$$

where the canonical kinetic energy is given by $\mathcal{G}(X) = X = -\frac{1}{2}\nabla_\mu\phi\nabla^\mu\phi$. K-essence was originally proposed as a model for inflation, and then as a model for dark energy, along with explorations of unifying dark energy and dark matter [5, 8, 9]. Another motivations to consider this type of lagrangian originates from string theory [10].

In this framework, gravitational and matter variables have been reduced to a finite number of degrees of freedom. For homogenous cosmological models the metric depends only on time and gives a model with a finite dimensional configuration space, called minisuperspace. In this work, we use this formulation to obtain classical solutions to the anisotropic Bianchi type I cosmological model with a perfect fluid. This class of models were considered initially in this

Journal of Physics: Conference Series **545** (2014) 012015 doi:10.1088/1742-6596/545/1/012015

formalism by Chimento and Forte [11]. The first step is to write the theory for the Bianchi type I model in the usual manner, that is, we calculate the corresponding energy-momentum tensor to the scalar field and give the equivalent Lagrangian density. Next, by means of a Legendre transformation, we proceed to obtain the canonical Lagrangian $\mathcal{L}_{can}$, from which the classical Hamiltonian $\mathcal{H}$ can be found.

This work is arranged as follows. In section 2 we obtain the corresponding K-essence field equations and in simple way we applied to Cosmological Bianchi Class A models. In section 3 we construct the lagrangian and hamiltonian densities for the anisotropic Bianchi type I cosmological model. In section 4 we present some ideas in as the anisotropic cosmological model can obtain its isotropization via the mean volume function and next we obtain the classical exact solution for all values in the gamma parameter. Finally, section 5 is devoted to some final remarks.

2. K-essence field equation

One of the simplest K-essence models, without self interaction has the following lagrangian density

$$\mathcal{L}_{\rm geo} = \mathrm{R} + \mathrm{f}(\phi)\mathcal{G}(\mathrm{X}), \tag{2}$$

R is the scalar curvature, and $f(\phi)$ is an arbitrary function of the scalar field. From the Lagrangian (2) we can build the complete action

$$\mathrm{I} = \int_\Sigma \sqrt{-\mathrm{g}}(\mathcal{L}_{\rm geo} + \mathcal{L}_\Lambda + \mathcal{L}_{\rm mat})\mathrm{d}^4\mathrm{x}, \tag{3}$$

where $\mathcal{L}_{mat}$ is the matter Lagrangian, $\mathcal{L}_\Lambda = 2\Lambda$ is the cosmological constant lagrangian, and g is the determinant of the metric tensor. The field equations for this theory are

$$\begin{aligned} \mathrm{G}_{\alpha\beta} + \Lambda \mathrm{g}_{\alpha\beta} + \mathrm{f}(\phi)\left[\mathcal{G}_\mathrm{X}\phi_{,\alpha}\phi_{,\beta} + \mathcal{G}\mathrm{g}_{\alpha\beta}\right] &= -\mathrm{T}_{\alpha\beta}, & (4)\\ \mathrm{f}(\phi)\left[\mathcal{G}_\mathrm{X}\phi^{;\beta}_{;\beta} + \mathcal{G}_\mathrm{XX}\mathrm{X}_{;\beta}\phi^{;\beta}\right] + \frac{\mathrm{df}}{\mathrm{d}\phi}\left[\mathcal{G} - 2\mathrm{X}\mathcal{G}_\mathrm{X}\right] &= 0, & (5)\end{aligned}$$

where we work in units with $8\pi G = 1$ and, as usual, the semicolon means a covariant derivative and a subscripted X denotes differentiation with respect to X.

The same set of equations(4,5) is obtained if we consider the scalar field $X(\phi)$ as part of the matter content, i.e. say $\mathcal{L}_{\mathrm{X},\phi} = \mathrm{f}(\phi)\mathcal{G}(\mathrm{X})$ with the corresponding energy-momentum tensor

$$\mathcal{T}_{\alpha\beta} = \mathrm{f}(\phi)\left[\mathcal{G}_\mathrm{X}\phi_{,\alpha}\phi_{,\beta} + \mathcal{G}(\mathrm{X})\mathrm{g}_{\alpha\beta}\right]. \tag{6}$$

Considering the energy-momentum tensor of a barotropic perfect fluid, $\mathrm{T}_{\alpha\beta} = (\rho + \mathrm{P})\mathrm{u}_\alpha \mathrm{u}_\beta + \mathrm{P}\mathrm{g}_{\alpha\beta}$, with u_α the four-velocity, which satisfy the relation $\mathrm{u}_\mu \mathrm{u}^\mu = -1$, ρ the energy density and P the pressure of the fluid. For simplicity we consider a comoving perfect fluid. The pressure, the energy density and the four-velocity corresponding to the energy-momentum tensor of the field X, become

$$\mathrm{P(X)} = \mathrm{f}(\phi)\mathcal{G}, \qquad \rho(\mathrm{X}) = \mathrm{f}(\phi)\left[2\mathrm{X}\mathcal{G}_\mathrm{X} - \mathcal{G}\right], \qquad \mathrm{u}_\mu = \frac{\nabla_\mathrm{u}\phi}{\sqrt{2\mathrm{X}}}, \tag{7}$$

thus, the barotropic parameter is

$$\omega_\mathrm{X} = \frac{\mathcal{G}}{2\mathrm{X}\mathcal{G}_\mathrm{X} - \mathcal{G}}. \tag{8}$$

and we notice that the case of a constant barotropic index ω_X, (with the exception $\omega_X = 0$) can be obtained by the $\mathcal{G}$ function

$$\mathcal{G} = \mathrm{X}^{\frac{1+\omega_\mathrm{X}}{2\omega_\mathrm{X}}}. \tag{9}$$

We have the following states in the evolution of our universe in this formalism,

$$\begin{cases} \text{stiff matter}: & \omega_{\rm X} = 1, \rightarrow \mathcal{G}({\rm X}) = {\rm X}. \\ \text{Radiation:} & \omega_{\rm X} = \frac{1}{3}, \rightarrow \mathcal{G}({\rm X}) = {\rm X}^2. \\ \text{inflation like:} & \omega_{\rm X} = -\frac{1}{3}, \rightarrow \mathcal{G}({\rm X}) = \frac{1}{\rm X}. \\ & \\ & \omega_{\rm X} = -\frac{2}{3}, \rightarrow \mathcal{G}({\rm X}) = \frac{1}{\sqrt[4]{X}}. \end{cases} \tag{10}$$

In reference [4], the authors present the analysis to radiation era using dynamical systems obtaining bouncing solutions.

2.1. Anisotropic cosmological Bianchi Class A models, f(ϕ)=*constant*

Considering the cosmological anisotropic Bianchi Class A models with metric (20), the equation (5) in term of X, becomes (here and all where appear the $\prime$ means, $\prime = \frac{d}{d\tau} = \frac{d}{Ndt}$, with t the usual cosmic time)

$$\left[\mathcal{G}_{\rm X} + 2{\rm X}\mathcal{G}_{\rm XX}\right]{\rm X}' + 6\Omega'{\rm X}\mathcal{G}_{\rm X} = 0, \tag{11}$$

where the scalar function Ω in some sense parametrize the mean scale factor of a isotropic universe, with its corresponding solution

$${\rm X}\mathcal{G}^2_{\rm X} = \eta {\rm e}^{-6\Omega}. \tag{12}$$

with η a constant.

Note that equation (12) give us the possible solutions X(A), as a function of the scale factor and therefore the behavior of all physical properties of the k-essence (like ρ, P) , are completely determined by the function X and do not depend on the evolution of the other types of energy density.

2.2. quintessence like case: $\mathcal{G} = X$ *and* f(ϕ) $\neq$ constant

The field equations for this particular case are

$$\begin{aligned} {\rm G}_{\alpha\beta} + \Lambda {\rm g}_{\alpha\beta} + {\rm f}(\phi)\left(\phi_{,\alpha}\phi_{,\beta} - \frac{1}{2}{\rm g}_{\alpha\beta}\phi_{,\gamma}\phi^{,\gamma}\right) &= -{\rm T}_{\alpha\beta}, & (13)\\ 2{\rm f}(\phi)\phi^{,\alpha}_{;\alpha} + \frac{\rm df}{\rm d\phi}\phi_{,\gamma}\phi^{,\gamma} &= 0, & (14) \end{aligned}$$

and the energy-momentum tensor (6) has the following form,

$$\mathcal{T}_{\alpha\beta} = {\rm f}(\phi)\left(\phi_{,\alpha}\phi_{,\beta} - \frac{1}{2}{\rm g}_{\alpha\beta}\phi_{,\gamma}\phi^{,\gamma}\right). \tag{15}$$

In this new line of reasoning, the action (3) can be rewritten as a geometrical part and matter content (usual matter plus a term that corresponds to the exotic scalar field component of the K-essence theory). The equation of motion for the field ϕ (14) has the following property, using the metric of the Bianchi type I model (however, this is satisfied by all cosmological Bianchi Class A models),

$$3\Omega'\phi'{\rm f} + \phi''{\rm f} + \frac{1}{2}\frac{\rm df}{\rm d\phi}\phi'^2 = 0, \tag{16}$$

which can be integrated at once with the following result,

$$\frac{1}{2}{\rm f}(\phi)\,\phi'^2 = \eta{\rm e}^{-6\Omega}, \qquad \rightarrow \qquad \int\sqrt{{\rm f}(\phi)}{\rm d}\phi = \sqrt{2\eta}\int {\rm e}^{-3\Omega(\tau)}{\rm d}\tau. \tag{17}$$

here η is an integration constant and has the same sign as $f(\phi)$. Considering the particular form of $\mathrm{f}(\phi)=\omega\phi^{\mathrm{m}}$ or $\mathrm{f}(\phi)=\omega e^{\mathrm{m}\phi}$ with m and ω constants, the classical solutions for the field ϕ in quadrature are

$$\phi(\tau)=\begin{cases}\left[(\mathrm{m}+2)\sqrt{\frac{\eta}{2\omega}}\int \mathrm{e}^{-3\Omega}\mathrm{d}\tau\right]^{\frac{2}{\mathrm{m}+2}}, & f(\phi)=\omega\phi^{m}, \quad m\neq -2\\ \mathrm{Exp}\left\{\sqrt{\frac{2\eta}{\omega}}\int \mathrm{e}^{-3\Omega}\mathrm{d}\tau\right\}, & f(\phi)=\omega\phi^{-2}, \quad m=-2\\ \frac{2}{\mathrm{m}}\mathrm{Ln}\left[\mathrm{m}\sqrt{\frac{\eta}{2\omega}}\int \mathrm{e}^{-3\Omega}\mathrm{d}\tau\right], & f(\phi)=\omega e^{m\phi}, \quad m\neq 0\\ \sqrt{2\eta}\int \mathrm{e}^{-3\Omega}\mathrm{d}\tau & f(\phi)=\omega, \quad m=0\end{cases} \tag{18}$$

In our particular case, it is evident that the contribution of the scalar field is equivalent to a stiff fluid with a barotropic equation of state $\gamma=1$. This is an instance of the results of the analysis of the energy momentum tensor of a scalar field (15) by Madsen [12] for General Relativity with scalar matter and by Pimentel [13] for the general scalar tensor theory. In both works a free scalar field is equivalent to a stiff matter fluid. In this way, we write the action (3) in the usual form

$$\mathrm{I}=\int_{\Sigma}\sqrt{-\mathrm{g}}\,(\mathrm{R}+\mathcal{L}_{\Lambda}+\mathcal{L}_{\mathrm{mat}}+\mathcal{L}_{\phi})\,\mathrm{d}^4\mathrm{x}, \tag{19}$$

and consequently, the classical equivalence between the two theories. We can infer that this correspondence is also satisfied in the quantum regime, so we can use this structure for the quantization program, where the ADM formalism is well known for different classes of matter [14].

3. Hamiltonian for the Bianchi type I cosmological model

Let us recall here the canonical formulation in the ADM formalism of the diagonal Bianchi Class A models. The metric has the form

$$\mathrm{ds}^2=-(\mathrm{Ndt})^2+\mathrm{e}^{2\Omega(\mathrm{t})}\,(\mathrm{e}^{2\beta(\mathrm{t})})_{\mathrm{ij}}\,\omega^{\mathrm{i}}\,\omega^{\mathrm{j}}=-\mathrm{d}\tau^2+\mathrm{e}^{2\Omega(\mathrm{t})}\,(\mathrm{e}^{2\beta(\mathrm{t})})_{\mathrm{ij}}\,\omega^{\mathrm{i}}\,\omega^{\mathrm{j}}, \tag{20}$$

where $\Omega(t)$ is a scalar, N the lapse function and $\beta_{\mathrm{ij}}(\mathrm{t})$ a 3x3 diagonal matrix, $\beta_{\mathrm{ij}}=\mathrm{diag}(\beta_{+}+\sqrt{3}\beta_{-},\beta_{+}-\sqrt{3}\beta_{-},-2\beta_{+})$, ω^{i} are one-forms that characterize each cosmological Bianchi type model and obey $\mathrm{d}\omega^{\mathrm{i}}=\frac{1}{2}\mathrm{C}^{\mathrm{i}}_{\mathrm{jk}}\omega^{\mathrm{j}}\wedge\omega^{\mathrm{k}}$, $\mathrm{C}^{\mathrm{i}}_{\mathrm{jk}}$ the structure constants of the corresponding invariance group. For the Bianchi type I model we have

$$\omega^1=\mathrm{dx}^1 \qquad \omega^2=\mathrm{dx}^2, \qquad \omega^3=\mathrm{dx}^3$$

The total Lagrangian density then for this metric becomes

$$\mathcal{L}_{\mathrm{I}}=\mathrm{e}^{3\Omega}\left[6\frac{\dot{\Omega}^2}{\mathrm{N}}-6\frac{\dot{\beta}_{+}^2}{\mathrm{N}}-6\frac{\dot{\beta}_{-}^2}{\mathrm{N}}+\frac{\mathrm{f}(\phi)}{2\mathrm{N}}\dot{\phi}^2+2\mathrm{N}\rho+2\mathrm{N}\Lambda\right], \tag{21}$$

and the corresponding Hamiltonian density is

$$\mathcal{H}_{\perp}=\frac{\mathrm{e}^{-3\Omega}}{24}\left(-\Pi_{\Omega}^2-\frac{12}{\mathrm{f}(\phi)}\Pi_{\phi}^2+\Pi_{+}^2+\Pi_{-}^2+\mathrm{b}_{\gamma}\mathrm{e}^{-3(\gamma-1)\Omega}+48\Lambda\mathrm{e}^{6\Omega}\right), \tag{22}$$

with $\mathrm{b}_{\gamma}=48\mu_{\gamma}$, where we have used energy-momentum conservation the law for a perfect fluid, $\mathrm{T}^{\mu\nu}_{;\nu}=0,\rightarrow\rho=\mu_{\gamma}\mathrm{e}^{-3(\gamma+1)\Omega}$, we assumed an equation of state $\mathrm{p}=\gamma\rho$,

Journal of Physics: Conference Series **545** (2014) 012015 doi:10.1088/1742-6596/545/1/012015

3.1. Classical equations

The corresponding Einstein field equations (13) and (14) for the anisotropic cosmological model bianchi type I are the following (remember that the prime $'$ is the derivative over the time $d\tau = \mathrm{N}dt$,

$$\begin{aligned}
&3\Omega'^2 - 3\beta_+'^2 - 3\beta_-'^2 - \frac{f}{4}\phi'^2 - \rho - \Lambda = 0\\
&2\Omega'' + 3\Omega'^2 - 3\Omega'\beta_+' - 3\sqrt{3}\Omega'\beta_-' - \beta_+'' + 3\beta_+'^2 - \sqrt{3}\beta_-'' + 3\beta_-'^2 + \frac{\mathrm{f}}{4}\phi'^2 + \mathrm{p} - \Lambda = 0, \quad (23)\\
&2\Omega'' + 3\Omega'^2 - 3\Omega'\beta_+' + 3\sqrt{3}\Omega'\beta_-' - \beta_+'' + 3\beta_+'^2 + \sqrt{3}\beta_-'' + 3\beta_-'^2 + \frac{\mathrm{f}}{4}\phi'^2 + \mathrm{p} - \Lambda = 0,\\
&2\Omega'' + 3\Omega'^2 + 6\Omega'\beta_+' + 2\beta_+'' + 3\beta_+'^2 + 3\beta_-'^2 + \frac{\mathrm{f}}{4}\phi'^2 + \mathrm{p} - \Lambda = 0\\
&\mathrm{f}\left(3\Omega'\phi' + \phi''\right) + \frac{1}{2}\frac{\mathrm{df}}{\mathrm{d}\phi}\phi'^2 = 0, \quad (24)
\end{aligned}$$

the solution of this last equation was putted in (17),

$$\frac{1}{2}\mathrm{f}(\phi)\,\phi'^2 = \eta \mathrm{e}^{-6\Omega}, \qquad \rightarrow \qquad \int \sqrt{\mathrm{f}(\phi)}\mathrm{d}\phi = \sqrt{2\eta}\int \mathrm{e}^{-3\Omega(\tau)}\mathrm{d}\tau. \quad (25)$$

The combination between the second and third equations us give the solution for the anisotropic function β_-, also the sum of third and fourth equations, putting the β_- solution, give us the form of the β_+ function,

$$\beta_\pm(\tau) = \mathrm{a}_\pm \int \mathrm{e}^{-3\Omega(\tau)}\mathrm{d}\tau, \quad (26)$$

where $\mathrm{a}_\pm$ are integration constants. So, the (23) are rewritten as

$$\begin{aligned}
3\Omega'^2 &= 3\mathrm{c}_1\mathrm{e}^{-6\Omega} + \mu_\gamma \mathrm{e}^{-3(\gamma+1)\Omega} + \Lambda, \qquad \mathrm{c}_1 = \mathrm{a}_+^2 + \mathrm{a}_-^2 + \frac{\eta}{6},\\
&2\Omega'' + 3\Omega'^2 + 3\beta_+'^2 + 3\beta_-'^2 + \frac{\mathrm{f}}{4}\phi'^2 + \mathrm{p} - \Lambda = 0, \quad (27)
\end{aligned}$$

4. Isotropization

The current observations of the cosmic background radiation set a very stringent limit to the anisotropy of the universe ([15]), therefore it is important to consider the anisotropy of the solutions. Recalling the Friedmann equation (constraint equation),

$$3\Omega'^2 - 3\beta_+'^2 - 3\beta_-'^2 - \frac{f}{4}\phi'^2 - \rho - \Lambda = 0, \quad (28)$$

we can see that isotropization is achieved when the terms with $\beta_\pm'^2$ go to zero or are negligible with respect to the other terms in the differential equation. We find in the literature the criteria for isotropization, among others, $(\beta_+'^2 + \beta_-'^2)/H^2 \rightarrow 0$, $(\beta_+'^2 + \beta_-'^2)/\rho \rightarrow 0$, that are consistent with our above remark. In the present case the comparison with the density should include the contribution of the scalar field. We define an anisotropic density ρ_a, that is proportional to the shear scalar,

$$\rho_a = \beta_+'^2 + \beta_-'^2, \quad (29)$$

and will compare it with ρ_γ,ρ_ϕ, and Ω'^2. From the Hamilton Jacobi analysis we now that

$$\rho_a \sim e^{-6\Omega}, \qquad \rho_\phi \sim e^{-6\Omega}, \qquad \Omega'^2 \sim 48\Lambda + {\kappa_\Omega}^2 e^{-6\Omega} + b_\gamma e^{-3(1+\gamma)\Omega} \quad (30)$$

and the ratios are

$$\frac{\rho_a}{\rho_\phi} \sim constant, \qquad \frac{\rho_a}{\rho_\gamma} \sim e^{3\Omega(\gamma-1)}, \qquad \frac{\rho_a}{\Omega'^2} \sim \frac{1}{\kappa_\Omega{}^2 + 48\Lambda e^{6\Omega} + b_\gamma e^{3(1-\gamma)\Omega}}. \tag{31}$$

Here we see that for expanding an universe the anisotropic density is dominated by the fluid density (with the exception of the stiff fluid) or by the Ω'^2 term and then at late times the isotropization is obtained if the expansion goes to infinity. Hence it is necessary to determine when we have an ever expanding universe.

In the following we obtain exact solutions in order to gives the volume function V to each case.

4.1. Exact Classical solutions

In order to find the solutions for the remaining minisuperspace variables we employ the Einstein-Hamilton-Jacobi equation, which arises by making the identification $\frac{\partial S(\Omega,\beta_\pm,\phi)}{\partial q^\mu} = \Pi_\mu$ in the Hamiltonian constraint $\mathcal{H}_\perp = 0$, which results in

$$\left(\frac{\partial S}{\partial \Omega}\right)^2 - \left(\frac{\partial S}{\partial \beta_+}\right)^2 - \left(\frac{\partial S}{\partial \beta_-}\right)^2 + \frac{12}{f(\phi)}\left(\frac{\partial S}{\partial \phi}\right)^2 - b_\gamma e^{3\Omega(1-\gamma)} - 48\Lambda e^{6\Omega} = 0 \tag{32}$$

in order to solve the above equation, we assume a solution of the form $S(\Omega, \beta_\pm, \phi) = S_1(\Omega) + S_2(\beta_+) + S_3(\beta_-) + S_4(\phi)$ which results in the following set of ordinary differential equations

$$\left(\frac{dS_1}{d\Omega}\right)^2 - \left(b_\gamma e^{-3(\gamma-1)\Omega} + 48\Lambda e^{6\Omega} + \kappa_\Omega{}^2\right) = 0 \tag{33}$$

$$\left(\frac{dS_2}{d\beta_+}\right)^2 - \kappa_+{}^2 = 0 \tag{34}$$

$$\left(\frac{dS_3}{d\beta_-}\right)^2 - \kappa_-{}^2 = 0 \tag{35}$$

$$\frac{12}{f(\phi)}\left(\frac{dS_4}{d\phi}\right)^2 - \kappa_\phi{}^2 = 0. \tag{36}$$

Here the κ_i are separation constants satisfying the relation $\kappa_\Omega{}^2 = \kappa_+{}^2 + \kappa_-{}^2 + \kappa_\phi{}^2$, $\kappa_\pm$ are real, $\kappa_\phi{}^2$ should have the same signs as $f(\phi)$ and for consistency with Eq.(17) we have $\kappa_\phi{}^2 = 24\eta$. Recalling the expressions for the momenta we can obtain solutions for equations (33-36) in quadrature, in particular

$$\Delta\tau = 12\int \frac{d\Omega}{\sqrt{48\Lambda + \kappa_\Omega{}^2 e^{-6\Omega} + b_\gamma e^{-3(1+\gamma)\Omega}}}. \tag{37}$$

$$\Delta\beta_\pm = \mp\frac{\kappa_\pm}{12}\int e^{-3\Omega(\tau)}d\tau. \tag{38}$$

We already know the solution for (36). As can be seen from (38), in order to obtain solutions for $\beta_\pm$ one needs to find a solution for Ω, which can be obtained from (37), and this one does not have a general solution, however, it is possible find solutions for particular values of the barotropic parameter γ with $\Lambda \neq 0$.

(i) $\Lambda = 0$ and $\gamma \neq 1$
The equation (37) can be written as

$$d\tau = 12\frac{e^{3\Omega}d\Omega}{\sqrt{\kappa_\Omega{}^2 + b_\gamma e^{-3\Omega(\gamma-1)}}}. \tag{39}$$

when we consider the time transformations $d\tau = e^{3\gamma\Omega}dT$, and the change of variable $u = \kappa\Omega^2 + b_\gamma e^{-3(\gamma-1)\Omega}$, this equation has the solution

$$\Omega(T) = \mathrm{Ln}\left[\theta_\gamma T^2 + \delta_\gamma T\right]^{-\frac{1}{3(\gamma-1)}}, \tag{40}$$

where $\theta_\gamma = \left(\frac{\gamma-1}{8}\right)^2 b_\gamma$ and $\delta_\gamma = -\sqrt{\kappa\Omega^2}\frac{\gamma-1}{4}$. With this, the time transformation becomes

$$d\tau = \left[\theta_\gamma T^2 + \delta_\gamma T\right]^{-\frac{\gamma}{(\gamma-1)}} dT.$$

and the closed form is [16],

$$\tau = \frac{(1-\gamma)}{\delta_\gamma}\left[\theta_\gamma T^2 + \delta_\gamma T\right]^{\frac{1}{1-\gamma}} {}_2F_1\left(1, -\frac{2}{\gamma-1}; \frac{\gamma-2}{\gamma-1}; -\frac{T\theta_\gamma}{\delta_\gamma}\right), \tag{41}$$

here ${}_2F_1$ is a hypergeometric function. We also have

$$\int e^{-3\Omega} d\tau = \frac{1}{\delta_\gamma}\mathrm{Ln}\left[\frac{T}{\theta_\gamma T + \delta_\gamma}\right]. \tag{42}$$

The anisotropy functions and the scalar field are given by

$$\Delta\beta_\pm = \mp\frac{\kappa_\pm}{12\delta_\gamma}\mathrm{Ln}\left[\frac{T}{\theta_\gamma T + \delta_\gamma}\right]. \tag{43}$$

$$\phi(T) = \begin{cases} \left[(m+2)\sqrt{\frac{\eta}{2\omega}}\frac{1}{\delta_\gamma}\mathrm{Ln}\left[\frac{T}{\theta_\gamma T+\delta_\gamma}\right]\right]^{\frac{2}{m+2}}, & f(\phi) = \omega\phi^m, \quad m \neq -2 \\ \mathrm{Exp}\left\{\sqrt{\frac{2\eta}{\omega}}\frac{1}{\delta_\gamma}\mathrm{Ln}\left[\frac{T}{\theta_\gamma T+\delta_\gamma}\right]\right\}, & f(\phi) = \omega\phi^{-2}, \quad m = -2 \\ \frac{2}{m}\mathrm{Ln}\left[m\sqrt{\frac{\eta}{2\omega}}\frac{1}{\delta_\gamma}\mathrm{Ln}\left[\frac{T}{\theta_\gamma T+\delta_\gamma}\right]\right], & f(\phi) = \omega e^{m\phi}, \quad m \neq 0 \\ \sqrt{2\eta}\frac{1}{\delta_\gamma}\mathrm{Ln}\left[\frac{T}{\theta_\gamma T+\delta_\gamma}\right] & f(\phi) = \omega, \quad m = 0 \end{cases} \tag{44}$$

As a concrete example we consider the particular value $\gamma = 0$, then $\tau = T$ and Ω becomes

$$\Omega(\tau) = \mathrm{Ln}\left[\frac{3}{4}\mu_0 T^2 + \frac{\sqrt{\kappa\Omega^2}}{4}T\right]^{\frac{1}{3}}, \qquad \Rightarrow \qquad e^{3\Omega} = \frac{3}{4}\mu_0 T^2 + \frac{\sqrt{\kappa\Omega^2}}{4}T. \tag{45}$$

So, the classical solutions for the anisotropic function $\beta_\pm$ and ϕ field are,

$$\Delta\beta_\pm = \mp\frac{\kappa_\pm}{3\sqrt{\kappa\Omega^2}}\mathrm{Ln}\left[\frac{\tau}{\frac{\sqrt{\kappa\Omega^2}}{4} + \frac{3}{4}\mu_0\tau}\right]. \tag{46}$$

$$\phi(\tau) = \begin{cases} \left[(m+2)\sqrt{\frac{\eta}{2\omega}}\mathrm{Ln}\left[\frac{\tau}{\frac{\sqrt{\kappa\Omega^2}}{4}+\frac{3}{4}\mu_0\tau}\right]\right]^{\frac{2}{m+2}}, & f(\phi) = \omega\phi^m, \quad m \neq -2 \\ \mathrm{Exp}\left\{\sqrt{\frac{2\eta}{\omega}}\mathrm{Ln}\left[\frac{\tau}{\frac{\sqrt{\kappa\Omega^2}}{4}+\frac{3}{4}\mu_0\tau}\right]\right\}, & f(\phi) = \omega\phi^{-2}, \quad m = -2 \\ \frac{2}{m}\mathrm{Ln}\left[m\sqrt{\frac{\eta}{2\omega}}\mathrm{Ln}\left[\frac{\tau}{\frac{\sqrt{\kappa\Omega^2}}{4}+\frac{3}{4}\mu_0\tau}\right]\right], & f(\phi) = \omega e^{m\phi}, \quad m \neq 0 \\ \sqrt{2\eta}\frac{4}{\sqrt{\kappa\Omega^2}}Ln\left[\frac{\tau}{\frac{\sqrt{\kappa\Omega^2}}{4}+\frac{3}{4}\mu_0\tau}\right]. & f(\phi) = \omega, \quad m = 0 \end{cases} \tag{47}$$

(ii) $\Lambda = 0$ and $\gamma = 1$
In this case equation (37) is

$$\Delta\tau = \int \frac{12}{\sqrt{\mathrm{b_2 e^{-6\Omega}}}} d\Omega \tag{48}$$

with $\mathrm{b_2} = \kappa_\Omega{}^2 + 48\mu_1$ that we assume positive.
The corresponding solutions are

$$\begin{aligned}
e^{3\Omega} &= \frac{\sqrt{\mathrm{b_2}}\Delta\tau}{4}, \\
\Delta\beta_\pm &= \mp\frac{\kappa_\pm}{12}\int e^{-3\Omega(\tau)}d\tau = \mp\kappa_\pm\frac{1}{3\sqrt{\mathrm{b_2}}}\mathrm{Ln}(\Delta\tau), \\
\phi(\tau) &= \begin{cases} \left[(\mathrm{m}+2)\sqrt{\frac{\eta}{2\omega}}\frac{4}{\sqrt{\mathrm{b_2}}}\mathrm{Ln}(\Delta\tau)\right]^{\frac{2}{\mathrm{m}+2}}, & f(\phi) = \omega\phi^m, \quad m \neq -2 \\ \mathrm{Exp}\left\{\sqrt{\frac{2\eta}{\omega}}\frac{4}{\sqrt{\mathrm{b_2}}}\mathrm{Ln}(\Delta\tau)\right\}, & f(\phi) = \omega\phi^{-2}, \quad m = -2 \\ \frac{2}{\mathrm{m}}\mathrm{Ln}\left[\mathrm{m}\sqrt{\frac{\eta}{2\omega}}\frac{4}{\sqrt{\mathrm{b_2}}}\mathrm{Ln}(\Delta\tau)\right], & f(\phi) = \omega e^{m\phi}, \quad m \neq 0 \\ \sqrt{2\eta}\frac{4}{\sqrt{\mathrm{b_2}}}\mathrm{Ln}(\Delta\tau), & f(\phi) = \omega, \quad m = 0 \end{cases}
\end{aligned} \tag{49}$$

(iii) $\Lambda \neq 0$ and $\gamma = -1$
(37) has the form

$$\Delta\tau = \int \frac{12}{\sqrt{\kappa_\Omega{}^2 e^{-6\Omega} + \mathrm{b_3}}} d\Omega \tag{50}$$

where $\mathrm{b_3} = 48\mu_{-1} + 48\Lambda$.

$$\Delta\tau = \frac{4}{\sqrt{\mathrm{b_3}}}\,\mathrm{arccsch}\left(\sqrt{\frac{\kappa_\Omega{}^2}{\mathrm{b_3}}}e^{-3\Omega}\right) \tag{51}$$

solving for Ω

$$\Omega = \frac{1}{3}\mathrm{Ln}\left|\sqrt{\frac{\kappa_\Omega{}^2}{\mathrm{b_3}}}\,\sinh\left(\frac{\sqrt{\mathrm{b_3}}}{4}\Delta\tau\right)\right|. \tag{52}$$

So, the others solutions become

$$\Delta\beta_\pm(\tau) = \pm\frac{\kappa_\pm}{3\sqrt{\kappa_\Omega{}^2}}\,\mathrm{Ln}\,|\Sigma|, \qquad \Sigma = \tanh\left(\frac{\sqrt{\mathrm{b_3}}}{4}\Delta\tau\right) \tag{53}$$

$$\phi(\tau) = \begin{cases} \left[-(\mathrm{m}+2)\sqrt{\frac{\eta}{2\omega}}\frac{4}{\sqrt{\kappa_\Omega{}^2}}\mathrm{Ln}\,|\Sigma|\right]^{\frac{2}{\mathrm{m}+2}}, & f(\phi) = \omega\phi^m, \quad m \neq -2 \\ \mathrm{Exp}\left\{-\sqrt{\frac{2\eta}{\omega}}\frac{4}{\sqrt{\kappa_\Omega{}^2}}\mathrm{Ln}\,|\Sigma|\right\}, & f(\phi) = \omega\phi^{-2}, \quad m = -2 \\ \frac{2}{\mathrm{m}}\mathrm{Ln}\left[-\mathrm{m}\sqrt{\frac{\eta}{2\omega}}\frac{4}{\sqrt{\kappa_\Omega{}^2}}\mathrm{Ln}\,|\Sigma|\right], & f(\phi) = \omega e^{m\phi}, \quad m \neq 0 \\ -\sqrt{2\eta}\frac{4}{\sqrt{\kappa_\Omega{}^2}}\mathrm{Ln}\,|\Sigma| & f(\phi) = \omega, \quad m = 0 \end{cases} \tag{54}$$

with the condition $\omega > 0$.

Journal of Physics: Conference Series **545** (2014) 012015 doi:10.1088/1742-6596/545/1/012015

(iv) $\Lambda \neq 0$ and $\gamma = 0$
For this case, equation (37) is

$$\begin{aligned}\Delta\tau &= \int \frac{12\mathrm{e}^{3\Omega}}{\sqrt{\kappa_\Omega{}^2 + 48\mu_0 \mathrm{e}^{3\Omega} + 48\Lambda \mathrm{e}^{6\Omega}}} \mathrm{d}\Omega &(55)\\ &= \frac{1}{\sqrt{3\Lambda}} \mathrm{Ln}\left[\frac{\mathrm{b}_0 + 96\Lambda \mathrm{e}^{3\Omega}}{48\Lambda} + \frac{1}{2\sqrt{3\Lambda}}\sqrt{\kappa_\Omega{}^2 + \mathrm{b}_0 \mathrm{e}^{3\Omega} + 48\Lambda \mathrm{e}^{6\Omega}}\right] &(56)\end{aligned}$$

with $b_0 = 48\mu_0$, and $\Lambda > 0$.
The function Ω become

$$\Omega = \frac{1}{3}\mathrm{Ln}\left[\frac{12\Lambda\left(\mathrm{e}^{\sqrt{3\Lambda}\Delta\tau} - \frac{\mathrm{b}_0}{48\Lambda}\right)^2 - \kappa_\Omega{}^2}{48\Lambda \mathrm{e}^{\sqrt{3\Lambda}\Delta\tau}}\right] \tag{57}$$

and we have for the anisotropic functions $\beta_\pm$ and field ϕ,

$$\begin{aligned}\Delta\beta_\pm(\tau) &= \pm\frac{2\kappa_\pm}{3\sqrt{\kappa_\Omega{}^2}} \operatorname{arctanh}(\xi), \qquad \xi = 2\sqrt{\frac{3\Lambda}{\kappa_\Omega{}^2}}\left(-\frac{\mathrm{b}_0}{48\Lambda} + \mathrm{e}^{\sqrt{3\Lambda}\Delta\tau}\right)\\ \phi(\tau) &= \begin{cases} \left[-(\mathrm{m}+2)\sqrt{\frac{\eta}{2\omega}}\frac{8}{\sqrt{\kappa_\Omega{}^2}}\operatorname{arctanh}(\xi)\right]^{\frac{2}{\mathrm{m}+2}}, & f(\phi) = \omega\phi^m, \quad m \neq -2\\ \mathrm{Exp}\left\{-\sqrt{\frac{2\eta}{\omega}}\frac{8}{\sqrt{\kappa_\Omega{}^2}}\operatorname{arctanh}(\xi)\right\}, & f(\phi) = \omega\phi^{-2}, \quad m = -2\\ \frac{2}{\mathrm{m}}\mathrm{Ln}\left[-\mathrm{m}\sqrt{\frac{\eta}{2\omega}}\frac{8}{\sqrt{\kappa_\Omega{}^2}}\operatorname{arctanh}(\xi)\right], & f(\phi) = \omega e^{m\phi}, \quad m \neq 0\\ -\sqrt{2\eta}\frac{8}{\sqrt{\kappa_\Omega{}^2}}\operatorname{arctanh}(\xi) & f(\phi) = \omega, \quad m = 0\end{cases} \quad (58)\end{aligned}$$

(v) $\Lambda \neq 0$ and $\gamma = 1$
Equation (37) becomes

$$\Delta\tau = \int \frac{12}{\sqrt{\mathrm{b}_4 \mathrm{e}^{-6\Omega} + 48\Lambda}} \mathrm{d}\Omega \tag{59}$$

where $\mathrm{b}_4 = \kappa_\Omega{}^2 + 48\mu_1$. In this case also we have two possible solutions depending on the value of the cosmological constant

- $\Lambda > 0$.
  The solution become

$$\Delta\tau = \frac{1}{\sqrt{3\Lambda}} \operatorname{arcsinh}\left(4\sqrt{\frac{3\Lambda}{\mathrm{b}_4}}\mathrm{e}^{3\Omega}\right) \tag{60}$$

so, the function Ω is

$$\Omega = \frac{1}{3}\mathrm{Ln}\left|\frac{1}{4}\sqrt{\frac{\mathrm{b}_4}{3\Lambda}}\sinh\left(\sqrt{3\Lambda}\Delta\tau\right)\right|. \tag{61}$$

The others functions become

$$\begin{aligned}\Delta\beta_\pm &= \pm\frac{\kappa_\pm}{3\sqrt{\mathrm{b}_4}}\mathrm{Ln}\left[\Xi\right], \qquad \Xi = \tanh\left(\sqrt{3\Lambda}\Delta\tau\right)\\ \phi(\tau) &= \begin{cases} \left[-(\mathrm{m}+2)\sqrt{\frac{\eta}{2\omega}}\frac{4}{\sqrt{\mathrm{b}_4}}\mathrm{Ln}\left[\Xi\right]\right]^{\frac{2}{\mathrm{m}+2}}, & f(\phi) = \omega\phi^m, \quad m \neq -2\\ \mathrm{Exp}\left\{-\sqrt{\frac{2\eta}{\omega}}\frac{4}{\sqrt{\mathrm{b}_4}}\mathrm{Ln}\left[\Xi\right]\right\}, & f(\phi) = \omega\phi^{-2}, \quad m = -2\\ \frac{2}{\mathrm{m}}\mathrm{Ln}\left[-\mathrm{m}\sqrt{\frac{\omega\eta}{2\omega}}\frac{4}{\sqrt{\mathrm{b}_4}}\mathrm{Ln}\left[\Xi\right]\right], & f(\phi) = \omega e^{m\phi}, \quad m \neq 0\\ -\sqrt{2\eta}\frac{4}{\sqrt{\mathrm{b}_4}}\mathrm{Ln}\left[\Xi\right] & f(\phi) = \omega, \quad m = 0\end{cases} \quad (62)\end{aligned}$$

- $\Lambda < 0$.
  The corresponding solutions are

$$\Delta\tau = -\frac{1}{\sqrt{3|\Lambda|}} \arccos\left(4\sqrt{\frac{3|\Lambda|}{\mathrm{b}_4}}\,\mathrm{e}^{3\Omega}\right) \tag{63}$$

the function Ω

$$\Omega = \frac{1}{3}\mathrm{Ln}\left|\frac{1}{4}\sqrt{\frac{\mathrm{b}_4}{3|\Lambda|}}\cos\left(\sqrt{3|\Lambda|}\Delta\tau\right)\right| \tag{64}$$

as the volume has an oscillatory behavior, the isotropization do not yield for this case, and for completeness we calculate the anisotropic functions $\beta_\pm$ and field ϕ,

$$\begin{aligned}
\Delta\beta_\pm &= \mp\frac{\kappa_\pm}{3\sqrt{\mathrm{b}_4}}\mathrm{Ln}\,|\psi|, \qquad \psi = \sec\left(\sqrt{3|\Lambda|}\,\Delta\tau\right) + \tan\left(\sqrt{3|\Lambda|}\,\Delta\tau\right),\\
\phi(\tau) &= \begin{cases}
\left[(\mathrm{m}+2)\sqrt{\frac{\eta}{2\omega}}\frac{4}{\sqrt{\mathrm{b}_4}}\mathrm{Ln}\,|\psi|\right]^{\frac{2}{\mathrm{m}+2}}, & f(\phi) = \omega\phi^m, \quad m \neq -2\\
\mathrm{Exp}\left\{\sqrt{\frac{2\eta}{\omega}}\frac{4}{\sqrt{\mathrm{b}_4}}\mathrm{Ln}\,|\psi|\right\}, & f(\phi) = \omega\phi^{-2}, \quad m = -2\\
\frac{2}{\mathrm{m}}\,\mathrm{Ln}\left[\mathrm{m}\sqrt{\frac{\eta}{2\omega}}\frac{4}{\sqrt{\mathrm{b}_4}}\mathrm{Ln}\,|\psi|\right], & f(\phi) = \omega e^{m\phi}, \quad m \neq 0\\
\sqrt{2\eta}\frac{4}{\sqrt{\mathrm{b}_4}}\mathrm{Ln}\,|\psi| & f(\phi) = \omega, \quad m = 0
\end{cases}
\end{aligned} \tag{65}$$

5. Final remarks

In this work we present the study of the classical cosmological anisotropic Bianchi type I in the K-essence formalism. In previous work made by Chimento and co-researcher [11], they present the possible isotropization of this model. Our goal in this work is that we obtain the corresponding classical solutions for a barotropic perfect fluid and cosmological term Λ that mimic the scalar field in equation (1). In the case of $\Lambda = 0$ and $\gamma \neq 1$ we obtain the solutions in closed form. With these solutions we can validate our qualitative analysis on isotropization of the cosmological model, implying that these become isotropic when the volume is large in the corresponding *time* evolution. So, only one solutions do not present the large volume, when $\Lambda < 0$ in stiff matter era in the ordinary matter content.

Acknowledgments

This work was partially supported by CONACYT 167335, 179881 grants. PROMEP grants UGTO-CA-3 and UAM-I-43. This work is part of the collaboration within the Instituto Avanzado de Cosmología and Red PROMEP: Gravitation and Mathematical Physics under project *Quantum aspects of gravity in cosmological models, phenomenology and geometry of space-time.* Many calculations where done by Symbolic Program REDUCE 3.8.

[1] Armendariz-Picon C, Mukhanov V and Steinbardt P J 2000 *Phys. Lett.* **85**, 4438 ; 2001 *Phys. Rev. D* **63**, 103510
[2] Bose N and Majumdar A S 2009 *Phys. Rev. D* **79**, 103517 *A k-essence model of inflation, dark matter and dark energy*, [arXiv:0812.4131]
[3] De-Santiago J and Jorge L. Cervantes-Cota 2011 *Phys. Rev. D* **83**, 063502 *Generalizing a Unified Model of Dark Matter, Dark Energy, and Inflation with Non Canonical Kinetic Term*
[4] De-Santiago J, Jorge L. Cervantes-Cota, David Wands 2013 *Phys. Rev. D* **87**, 023502 *Cosmological phase space analysis of the $F(X) - V(\phi)$ scalar field and bouncing solutions*
[5] Roland de Putter and Eric V. Linder 2007 *Astropart. Phys.* **28**, 263 *Kinetic k-essence and Quintessence.* [arXiv:0705.0400]
[6] Chiba T, Dutta S and Scherrer R J 2009 *Phys. Rev. D* **80**, 043517 *Slow-roll k-essence,* [arXiv:0906.0628]

[7] Arroja F and Sasaki M 2010 *Phys. Rev. D* **81**, 107301 *A note on the equivalence of a barotropic perfect fluid with a k-essence scalar field,* [arXiv:1002.1376]
[8] Bilic N, Tupper G and Viollier R 2002 *Phys.Lett. B* **535**, 17
[9] Bento M, Bertolami O and Sen A 2002 *Phys. Rev. D* **66**, 043507
[10] Armendariz-Picon C, Damour T and Mukhanov V 1999 *Phys. Lett. B* **458**, 209; Garriga J and Mukhanov C 1999 *Phys. Lett. B* **458**, 219
[11] Luis P. Chimento and Mónica Forte 2006 *Phys. Rev. D* **73**, 063502 *Anisotropic k-essence cosmologies*
[12] Madsen M 1988 *Class. and Quantum Grav.* **5**, 627
[13] Pimentel L O 1989 *Class. and Quantum Grav.* **6**, L263
[14] Ryan M P 1972 *Hamiltonian cosmology,* (Springer, Berlin)
[15] Martinez-Gonzalez E and Sanz J L 1995 Astron. Astrophys. 300 346
[16] Andrei C Polyanin and Valentin F Zaitsev 2003 *Handbook of Exact solutions for ordinary differential equations,* (Second edition, Chapman & Hall/CRC)

Journal of Physics: Conference Series **545** (2014) 012016 doi:10.1088/1742-6596/545/1/012016

Supersymmetric closed string tachyon cosmology: a first approach

V Vázquez-Báez[1,2] and C Ramírez[2]

[1]Benemérita Universidad Autónoma de Puebla, Facultad de Ingeniería, Blvd. Valsequillo esq. Av. San Claudio s/n, Ciudad Universitaria, Edificio 108-C, 72570 Puebla, México
[2]Benemérita Universidad Autónoma de Puebla, Facultad de Ciencias Físico Matemáticas, P.O. Box 165, 72000 Puebla, México

E-mail: manuel.vazquez@correo.buap.mx

Abstract. We give a worldline supersymmetric formulation for the effective action of closed string tachyon in a FRW background. This is done considering that, as shown by Vafa, the effective theory of closed string tachyons can have worldsheet supersymmetry. The Hamiltonian is constructed by means of the Dirac procedure and written in a quantum version. By using the supersymmetry algebra we are able to find solutions to the Wheeler-DeWitt equation via a more simple set of first order differential equations.

1. Introduction

String theory has in its lowest mode tachyons which for the lack of knowledge on how to handle unstable configurations were ignored for many years. With the inclusion of supersymmetry they can be consistently eliminated by the GSO truncation. Some years ago it has been seen that the evolution of these tachyonic instabilities can be described by the condensation of the tachyonic modes. This was first perfomed in the somewhat simpler case of open strings, resumed by the well known Sen conjectures [1]. For closed strings the situation is more complicated, because it involves the structure of space-time. An interesting fact in this case, is that closed string tachyons, which are nonsupersymmetric in target space, can have worldsheet supersymmetry [2]. In [3] a Lagrangian of supersymmetric tachyons in the framework of a FRW background has been worked out, in a worldline superspace where the time variable is extended to the superspace of supersymmetry [6, 7]. This work is performed along such approach considering the covariant formulation of one-dimensional supergravity of the so called 'new' Θ variables [8, 9], which allows in a straightforward way to write supergravity invariant actions. Supersymmetric cosmology was studied in a variety of different schemes, we refer the interested reader to the well known books by D'Eath and Moniz for a review [4, 5],

2. Closed String Tachyon Effective Action

The bosonic closed string tachyon effective action is given according to [10] as

$$S = \frac{1}{2\kappa_D^2}\int\sqrt{-g}e^{-2\phi}\left[R+4\left(\partial\phi\right)^2-\left(\partial T\right)^2-2V(T)\right]d^Dx, \qquad (1)$$

where T is the closed string tachyon field, $V(T)$ is the tachyon potential and ϕ is the dilaton field. This action can be written in the Einstein frame by means of $g_{\mu\nu}^{string} = e^{\phi} g_{\mu\nu}^{Einstein}$, which is more suitable for our cosmological approach. For a four dimensional FRW metric and in the Einstein frame, the action takes the form

$$S = \int \left[-\frac{3\dot{a}^2 a}{\kappa^2 N} + \frac{3Nka}{\kappa^2} + \frac{a^3}{\kappa^2 N}\dot{\phi}^2 + \frac{a^3}{2\kappa^2 N}\dot{T}^2 - \frac{a^3 N e^{2\phi} V(T)}{\kappa^2} \right] dt, \qquad (2)$$

where N is the lapse function and a is the scale factor. This Lagrangian is invariant under time reparametrizations and this invariance is extended to supersymmetry introducing a Grassmann superspace associated to the bosonic time coordinate t (see Tkach et al. in [6, 7]).

3. Supersymetric Closed String Tachyon Model

As stated in [11] superspace is the natural framework for a geometrical formulation of supersymmetry and supergravity, it extends spacetime by anticommuting Grassmann variables, $x^m \to (x^m, \theta^\mu)$, and the field content of the superfields is given by the Grassmann power expansion in the anticommuting variables $\phi(z) = \sum_n 1/n!\, \theta^{\mu_1} \cdots \theta^{\mu_n} \phi_{\mu_1 \cdots \mu_n}(x)$.

The supersymmetric cosmological model is obtained upon an extension of the time coordinate into a supermultiplet $t \to (t, \Theta, \bar{\Theta})$. Due to this generalization the fields of the theory are generalized as superfields, in this the expansion mentioned above is given by

$$\begin{aligned} \mathcal{A}\left(t,\Theta,\bar{\Theta}\right) &= a(t) + i\Theta\bar{\lambda}(t) + i\bar{\Theta}\lambda(t) + B(t)\Theta\bar{\Theta}, \\ \mathcal{T}\left(t,\Theta,\bar{\Theta}\right) &= T(t) + i\Theta\bar{\eta}(t) + i\bar{\Theta}\eta(t) + G(t)\Theta\bar{\Theta}, \\ \Phi\left(t,\Theta,\bar{\Theta}\right) &= \phi(t) + i\Theta\bar{\chi}(t) + i\bar{\Theta}\chi(t) + F(t)\Theta\bar{\Theta}, \end{aligned} \qquad (3)$$

where, $\mathcal{A}$, $\mathcal{T}$ and Φ are the superfields of a, T and ϕ.

The supersymmetric generalization of the action is given by

$$S = S_{Rsusy} + S_{Msusy}, \qquad (4)$$

where, S_{Rsusy} is the cosmological supersymmetric generalization of the free FRW model

$$S_{Rsusy} = \int \left(\frac{3\mathcal{E}}{\kappa^2} \mathcal{A} \nabla_{\bar{\theta}} \mathcal{A} \nabla_{\theta} \mathcal{A} - \frac{3\sqrt{k}}{\kappa^2} \mathcal{E}\mathcal{A}^2 \right) d\Theta d\bar{\Theta} dt, \qquad (5)$$

and the supersymmetric matter term is

$$S_{Msusy} = \frac{1}{\kappa^2} \int \left[-\mathcal{E}\mathcal{A}^3 \nabla_{\bar{\Theta}} \Phi \nabla_{\Theta} \Phi - \frac{1}{2}\mathcal{E}\mathcal{A}^3 \nabla_{\bar{\Theta}} \mathcal{T} \nabla_{\Theta} \mathcal{T} + \mathcal{E}\mathcal{A}^3 W(\Phi, \mathcal{T}) \right] d\Theta d\bar{\Theta} dt, \qquad (6)$$

where $W(\Phi, \mathcal{T})$ is the superpotential, and $\mathcal{E}$ is an invariant density playing the role of a supersymmetric $\sqrt{-g}$ in the action and given by

$$\mathcal{E} = -e - \frac{i}{2}(\Theta\bar{\Psi} + \bar{\Theta}\psi), \qquad (7)$$

see [3, 8, 9] for details of its calculation.

The superpotential expansion is $W(\Phi, \mathcal{T}) = W(\phi, T) + \frac{\partial W}{\partial \phi}(\Phi - \phi) + \frac{\partial W}{\partial T}(\mathcal{T} - T) + \frac{1}{2}\frac{\partial^2 W}{\partial T^2}(\mathcal{T} - T)^2 + \frac{1}{2}\frac{\partial^2 W}{\partial \phi^2}(\Phi - \phi)^2 + \frac{\partial^2 W}{\partial T \partial \phi}(\mathcal{T} - T)(\Phi - \phi)$. This expansion is finite because the terms $(\mathcal{T} - T)$ and $(\Phi - \phi)$ are purely grassmannian and nilpotency holds for these kind or variables.

Integrating over the Grassmann parameters, the total Lagrangian is

$$
\begin{aligned}
L = & -\frac{3a\dot{a}^2}{e\kappa^2} + \frac{3a\dot{a}}{e\kappa^2}\left(\psi\lambda - \bar{\psi}\bar{\lambda}\right) + \frac{a^3\dot{T}^2}{2e\kappa^2} - \frac{a^3\dot{T}}{2e\kappa^2}\left(\psi\eta - \bar{\psi}\bar{\eta}\right) + \frac{3ia^2\dot{T}}{2\kappa^2}\left(\lambda\bar{\eta} + \bar{\lambda}\eta\right) + \frac{a^3\dot{\phi}^2}{e\kappa^2} \\
& - \frac{a^3\dot{\phi}}{e\kappa^2}\left(\psi\chi - \bar{\psi}\bar{\chi}\right) + \frac{3ia^2\dot{\phi}}{\kappa^2}\left(\lambda\bar{\chi} + \bar{\lambda}\chi\right) + \frac{3ia}{\kappa^2}\left(\lambda\dot{\bar{\lambda}} + \bar{\lambda}\dot{\lambda}\right) - \frac{ia^3}{2\kappa^2}\left(\eta\dot{\bar{\eta}} + \bar{\eta}\dot{\eta}\right) - \frac{ia^3}{\kappa^2}\left(\chi\dot{\bar{\chi}} + \bar{\chi}\dot{\chi}\right) \\
& + \frac{6e\sqrt{k}\lambda\bar{\lambda}}{\kappa^2} + \frac{3ia\sqrt{k}}{\kappa^2}\left(\psi\lambda + \bar{\psi}\bar{\lambda}\right) - \frac{6aeW\lambda\bar{\lambda}}{\kappa^2} - \frac{3ia^2W}{2\kappa^2}\left(\psi\lambda + \bar{\psi}\bar{\lambda}\right) - \frac{ia^3W_T}{2\kappa^2}\left(\psi\eta + \bar{\psi}\bar{\eta}\right) \\
& + \frac{3a^2eW_T}{\kappa^2}\left(\bar{\lambda}\eta - \lambda\bar{\eta}\right) - \frac{ia^3W_\phi}{2\kappa^2}\left(\psi\chi + \bar{\psi}\bar{\chi}\right) + \frac{3a^2eW_\phi}{\kappa^2}\left(\bar{\lambda}\chi - \lambda\bar{\chi}\right) - \frac{a^3eW_{TT}}{\kappa^2}\eta\bar{\eta} \\
& + \frac{a^3eW_{T\phi}}{\kappa^2}\left(\bar{\chi}\eta - \chi\bar{\eta}\right) - \frac{a^3eW_{\phi\phi}}{\kappa^2}\chi\bar{\chi} - \frac{3a\psi\bar{\psi}\lambda\bar{\lambda}}{2e\kappa^2} + \frac{a^3\psi\bar{\psi}\eta\bar{\eta}}{4e\kappa^2} + \frac{a^3\chi\bar{\chi}\psi\bar{\psi}}{2e\kappa^2} - \frac{3ae\lambda\bar{\lambda}\eta\bar{\eta}}{\kappa^2} \\
& - \frac{6ae\lambda\bar{\lambda}\chi\bar{\chi}}{\kappa^2} - \frac{3aB^2e}{\kappa^2} + \frac{6aBe\sqrt{k}}{\kappa^2} - \frac{3Be\lambda\bar{\lambda}}{\kappa^2} - \frac{3a^2Be\eta\bar{\eta}}{2\kappa^2} - \frac{3a^2Be\chi\bar{\chi}}{\kappa^2} - \frac{3a^2BeW}{\kappa^2} \\
& + \frac{a^3eG^2}{2\kappa^2} + \frac{3a^2eG}{2\kappa^2}\left(\lambda\bar{\eta} - \bar{\lambda}\eta\right) - \frac{a^3eGW_T}{\kappa^2} + \frac{a^3eF^2}{\kappa^2} + \frac{3a^2eF}{\kappa^2}\left(\lambda\bar{\chi} - \bar{\lambda}\chi\right) - \frac{a^3eFW_\phi}{\kappa^2},
\end{aligned}
$$

where the subscripts in W denote partial differentiation with respect to ϕ and T respectively.

When we perform the variation of the Lagrangian with respect to the fields B, F and G, as usual algebraic constraints are obtained, that is B, F and G play the role of auxiliary fields, and they can be eliminated from the Lagrangian. Upon solving for the auxiliary fields and making the further rescalings for convinience $\lambda \to \kappa a^{-1/2}\lambda$, $\bar{\lambda} \to \kappa a^{-1/2}\bar{\lambda}$, $\eta \to \kappa a^{-3/2}\eta$, $\bar{\eta} \to \kappa a^{-3/2}\bar{\eta}$, $\chi \to \kappa a^{-3/2}\chi$, $\bar{\chi} \to \kappa a^{-3/2}\bar{\chi}$, we find the Lagrangian

$$
\begin{aligned}
L = & -\frac{3a\dot{a}^2}{e\kappa^2} + \frac{3\sqrt{a}\dot{a}}{e\kappa}\left(\psi\lambda - \bar{\psi}\bar{\lambda}\right) + \frac{3eka}{\kappa^2} + \frac{\dot{T}^2a^3}{2e\kappa^2} - \frac{\sqrt{a^3}\dot{T}}{2e\kappa}\left(\psi\eta - \bar{\psi}\bar{\eta}\right) + \frac{3i\dot{T}}{2}\left(\lambda\bar{\eta} + \bar{\lambda}\eta\right) \\
& + \frac{\dot{\phi}^2a^3}{e\kappa^2} - \frac{\sqrt{a^3}\dot{\phi}}{e\kappa}\left(\psi\chi - \bar{\psi}\bar{\chi}\right) + 3i\dot{\phi}\left(\lambda\bar{\chi} + \bar{\lambda}\chi\right) + 3i\left(\lambda\dot{\bar{\lambda}} + \bar{\lambda}\dot{\lambda}\right) - \frac{i}{2}\left(\eta\dot{\bar{\eta}} + \bar{\eta}\dot{\eta}\right) \\
& - i\left(\chi\dot{\bar{\chi}} + \bar{\chi}\dot{\chi}\right) + \frac{3e\sqrt{k}\lambda\bar{\lambda}}{a} - \frac{3e\sqrt{k}\eta\bar{\eta}}{2a} - \frac{3e\sqrt{k}\chi\bar{\chi}}{a} + \frac{3i\sqrt{ak}}{\kappa}\left(\psi\lambda + \bar{\psi}\bar{\lambda}\right) \\
& + \frac{3eW^2a^3}{4\kappa^2} - \frac{3e\sqrt{k}a^2W}{\kappa^2} - \frac{eW_T^2a^3}{2\kappa^2} - \frac{eW_\phi^2a^3}{4\kappa^2} - \frac{9}{2}eW\lambda\bar{\lambda} + \frac{3}{4}eW\eta\bar{\eta} + \frac{3}{2}eW\chi\bar{\chi} \\
& - \frac{3ia^{3/2}W}{2\kappa}\left(\psi\lambda + \bar{\psi}\bar{\lambda}\right) - \frac{i\sqrt{a^3}W_T}{2\kappa}\left(\psi\eta + \bar{\psi}\bar{\eta}\right) + \frac{3eW_T}{2}\left(\bar{\lambda}\eta - \lambda\bar{\eta}\right) - \frac{i\sqrt{a^3}W_\phi}{2\kappa}\left(\psi\chi + \bar{\psi}\bar{\chi}\right) \\
& + \frac{3eW_\phi}{2}\left(\bar{\lambda}\chi - \lambda\bar{\chi}\right) - eW_{TT}\eta\bar{\eta} + eW_{T\phi}\left(\bar{\chi}\eta - \chi\bar{\eta}\right) - eW_{\phi\phi}\chi\bar{\chi} + \frac{3e\kappa^2}{4a^3}\eta\bar{\eta}\chi\bar{\chi} - \frac{3}{2e}\psi\bar{\psi}\lambda\bar{\lambda} \\
& + \frac{1}{2e}\psi\bar{\psi}\chi\bar{\chi} + \frac{1}{4e}\psi\bar{\psi}\eta\bar{\eta}.
\end{aligned}
$$

Journal of Physics: Conference Series **545** (2014) 012016 doi:10.1088/1742-6596/545/1/012016

4. Hamiltonian analysis

We have the following canonical momenta

$$
\begin{aligned}
\pi_a &= -\frac{6a\dot{a}}{e\kappa^2} - \frac{3\sqrt{a}\bar{\psi}\lambda}{e\kappa} + \frac{3\sqrt{a}\psi\lambda}{e\kappa},\\
\pi_T &= \frac{a^3\dot{T}}{e\kappa^2} + \frac{\sqrt{a^3}\bar{\psi}\bar{\eta}}{2e\kappa} - \frac{\sqrt{a^3}\psi\eta}{2e\kappa} + \frac{3a^{3/2}i\lambda\bar{\eta}}{2\sqrt{a^3}} + \frac{3a^{3/2}i\bar{\lambda}\eta}{2\sqrt{a^3}},\\
\pi_\phi &= \frac{2a^3\dot{\phi}}{e\kappa^2} + \frac{\sqrt{a^3}\bar{\psi}\bar{\chi}}{e\kappa} - \frac{\sqrt{a^3}\psi\chi}{e\kappa} + \frac{3a^{3/2}i\lambda\bar{\chi}}{\sqrt{a^3}} + \frac{3a^{3/2}i\bar{\lambda}\chi}{\sqrt{a^3}},\\
\pi_\lambda &= -3i\bar{\lambda}, \quad \pi_{\bar{\lambda}} = -3i\lambda,\\
\pi_\eta &= \frac{i}{2}\bar{\eta}, \quad \pi_{\bar{\eta}} = \frac{i}{2}\eta,\\
\pi_\chi &= i\bar{\chi}, \quad \pi_{\bar{\chi}} = i\chi.
\end{aligned}
$$

It can be seen, as usual, the appearence of the fermionic constraints

$$
\begin{aligned}
\Omega_\lambda &= \pi_\lambda + 3i\bar{\lambda}, \quad \Omega_{\bar{\lambda}} = \pi_{\bar{\lambda}} + 3i\lambda,\\
\Omega_\eta &= \pi_\eta - \frac{i}{2}\bar{\eta}, \quad \Omega_{\bar{\eta}} = \pi_{\bar{\eta}} - \frac{i}{2}\eta,\\
\Omega_\chi &= \pi_\chi - i\bar{\chi}, \quad \Omega_{\bar{\chi}} = \pi_{\bar{\chi}} - i\chi.
\end{aligned} \tag{8}
$$

According to the Dirac formalism, the previous constraints are second class and the dynamics of the system is obtained when we impose the set of constraints (8) and introduce the Dirac brackets, we have in this case

$$
\begin{array}{c}
\{a,\pi_a\}_D = 1, \ \{\phi,\pi_\phi\}_D = 1, \ \{T,\pi_T\}_D = 1,\\
\{\lambda,\bar{\lambda}\}_D = -\frac{1}{6i}, \ \{\chi,\bar{\chi}\}_D = -\frac{i}{2}, \ \{\eta,\bar{\eta}\}_D = -i,
\end{array} \tag{9}
$$

as the only non zero brackets of the theory.

Following Dirac's procedure, using the standard definition for the Hamiltonian and imposing the constraints (8), the Hamiltonian of the system can be written as

$$
H = NH_0 + \frac{1}{2}\psi S - \frac{1}{2}\bar{\psi}\bar{S}, \tag{10}
$$

where

$$
\begin{aligned}
H_0 = &-\tfrac{\kappa^2\pi_a^2}{12a} + \tfrac{\kappa^2\pi_T^2}{2a^3} - \tfrac{3i\kappa^2\pi_T}{2a^3}\left(\lambda\bar{\eta} + \bar{\lambda}\eta\right) + \tfrac{\kappa^2\pi_\phi^2}{4a^3} - \tfrac{3i\kappa^2\pi_\phi}{2a^3}\left(\lambda\bar{\chi} + \bar{\lambda}\chi\right) - \tfrac{3a^3}{4\kappa^2}W^2\\
&+\tfrac{3\sqrt{k}a^2}{\kappa^2}W - \tfrac{3ka}{\kappa^2} + \tfrac{a^3}{2\kappa^2}W_T^2 + \tfrac{a^3}{4\kappa^2}W_\phi^2 + \tfrac{9}{2}W\lambda\bar{\lambda} - \tfrac{3}{4}W\eta\bar{\eta} - \tfrac{3}{2}W\chi\bar{\chi}\\
&+\tfrac{3}{2}W_T\left(\lambda\bar{\eta} - \bar{\lambda}\eta\right) + \tfrac{3}{2}W_\phi\left(\lambda\bar{\chi} - \bar{\lambda}\chi\right) + W_{TT}\eta\bar{\eta} + W_{T\phi}\left(\chi\bar{\eta} - \bar{\chi}\eta\right)\\
&+W_{\phi\phi}\chi\bar{\chi} - \tfrac{3\sqrt{k}}{a}\lambda\bar{\lambda} + \tfrac{3\sqrt{k}}{2a}\eta\bar{\eta} + \tfrac{3\sqrt{k}}{a}\chi\bar{\chi} - \tfrac{9\kappa^2}{2a^3}\lambda\bar{\lambda}\chi\bar{\chi} - \tfrac{9\kappa^2}{4a^3}\lambda\bar{\lambda}\eta\bar{\eta} - \tfrac{3\kappa^2}{4a^3}\eta\bar{\eta}\chi\bar{\chi},
\end{aligned} \tag{11}
$$

$$
\begin{aligned}
S = &\tfrac{\kappa\pi_a}{\sqrt{a}}\lambda + \tfrac{\kappa\pi_T}{\sqrt{a^3}}\eta + \tfrac{\kappa\pi_\phi}{\sqrt{a^3}}\chi - \tfrac{6i\sqrt{ak}}{\kappa}\lambda + \tfrac{3i\sqrt{a^3}}{\kappa}W\lambda\\
&+\tfrac{i\sqrt{a^3}}{\kappa}W_T\eta + \tfrac{i\sqrt{a^3}}{\kappa}W_\phi\chi + \tfrac{3i\kappa}{2a^{3/2}}\lambda\eta\bar{\eta} + \tfrac{3i\kappa}{a^{3/2}}\lambda\chi\bar{\chi},
\end{aligned} \tag{12}
$$

$$
\begin{aligned}
\bar{S} = &\tfrac{\kappa\pi_a}{\sqrt{a}}\bar{\lambda} + \tfrac{\kappa\pi_T}{\sqrt{a^3}}\bar{\eta} + \tfrac{\kappa\pi_\phi}{\sqrt{a^3}}\bar{\chi} + \tfrac{6i\sqrt{ak}}{\kappa}\bar{\lambda} - \tfrac{3i\sqrt{a^3}}{\kappa}W\bar{\lambda}\\
&-\tfrac{i\sqrt{a^3}}{\kappa}W_T\bar{\eta} - \tfrac{i\sqrt{a^3}}{\kappa}W_\phi\bar{\chi} - \tfrac{3i\kappa}{2a^{3/2}}\bar{\lambda}\eta\bar{\eta} - \tfrac{3i\kappa}{a^{3/2}}\bar{\lambda}\chi\bar{\chi},
\end{aligned} \tag{13}
$$

are the set of first class constraints satisfying the Dirac algebra $\{S,\bar{S}\}_D = 2H_0$, $\{H_0,S\}_D = \{H_0,\bar{S}\}_D = 0$.

5. Wave function of the universe

In a previews work, see for instance [12], we computed a wave function of the universe for a supersymmetric tachyon in a FRW backgraund, providing a normalizable solution in which the unstable nature of the tachyon can be apreciated; in such study the action (2) was used with the dilaton turned off. Now we address the solution of the Wheeler-DeWitt equation for the complete theory.

The constraints (11)-(13) become conditions on the wave function of the universe, i.e, $H\Psi = \bar{S}\Psi = S\Psi = F\Psi = 0$, which due to the operator superalgebra, specifically $\{S, \bar{S}\} = 2H$, imply that any state wave function, Ψ, satisfying $\bar{S}\Psi = S\Psi = 0$ is also solution of the Wheeler-DeWitt equation, $H\Psi = 0$.

There are bilinear terms in the constraints S and $\bar{S}$, eqs. (14) and (13), that need to be ordered to produce a suitable quantum theory, we make use of the anticommutation propeties of Grassmann variables at the clasical level to write such terms as commutators e.g. $\lambda\chi\bar{\chi} \to \lambda[\chi, \bar{\chi}]$. Thus we have for the quantum cosntraints:

$$S = \frac{\kappa\pi_a}{\sqrt{a}}\lambda + \frac{\kappa\pi_\phi}{\sqrt{a^3}}\chi + \frac{\kappa\pi_T}{\sqrt{a^3}}\eta + \frac{3i\sqrt{a^3}W}{\kappa}\lambda + \frac{i\sqrt{a^3}W_T}{\kappa}\eta + \frac{i\sqrt{a^3}W_\phi}{\kappa}\chi \\ + \frac{3i\kappa}{4a^{3/2}}\lambda[\eta, \bar{\eta}] + \frac{3i\kappa}{2a^{3/2}}\lambda[\chi, \bar{\chi}] - \frac{6i\sqrt{ak}}{\kappa}\lambda, \tag{14}$$

$$\bar{S} = \frac{\kappa\pi_a}{\sqrt{a}}\bar{\lambda} + \frac{\kappa\pi_\phi}{\sqrt{a^3}}\bar{\chi} + \frac{\kappa\pi_T}{\sqrt{a^3}}\bar{\eta} - \frac{3i\sqrt{a^3}W}{\kappa}\bar{\lambda} - \frac{i\sqrt{a^3}W_T}{\kappa}\bar{\eta} - \frac{i\sqrt{a^3}W_\phi}{\kappa}\bar{\chi} \\ - \frac{3i\kappa}{4a^{3/2}}\bar{\lambda}[\eta, \bar{\eta}] - \frac{3i\kappa}{2a^{3/2}}\bar{\lambda}[\chi, \bar{\chi}] + \frac{6i\sqrt{ak}}{\kappa}\bar{\lambda}. \tag{15}$$

We introduce a representation for the Grassmann variables in terms of suitable linear combinations of 8-dimension Dirac matrices, thus the wave function will be an eight-component spinor $\Psi = (\psi_1, \psi_2, \psi_3, \psi_4, \psi_5, \psi_6, \psi_7, \psi_8)$. Upon application of the operators S and $\bar{S}$ on this state, we get two sets of eight equations, from which we see that the only nonzero components are ψ_1 and ψ_6, which satisfy the following equations

$$\begin{aligned} \left(\hbar\kappa\frac{\partial}{\partial\phi} - \frac{a^3}{\kappa}\frac{\partial W}{\partial\phi}\right)\psi_1 = 0, \\ \left(\hbar\kappa\frac{\partial}{\partial T} - \frac{a^3}{\kappa}\frac{\partial W}{\partial T}\right)\psi_1 = 0, \\ \left(\hbar a\kappa^2\frac{\partial}{\partial a} - 3a^3W + 6a^2k + \frac{9}{8}\right)\psi_1 = 0, \end{aligned} \tag{16}$$

and

$$\begin{aligned} \left(\hbar\kappa^2\frac{\partial}{\partial\phi} + a^3\frac{\partial W}{\partial\phi}\right)\psi_6 = 0, \\ \left(\hbar\kappa\frac{\partial}{\partial T} + \frac{a^3}{\kappa}\frac{\partial W}{\partial T}\right)\psi_6 = 0, \\ \left(\hbar a\kappa^2\frac{\partial}{\partial a} + 3a^3W - 6a^2k + \frac{9}{8}\right)\psi_6 = 0. \end{aligned} \tag{17}$$

Solving these equations we find that

$$\psi_1(\phi, T, a) = a^{-\frac{9}{8\hbar}}\exp\left[\frac{a^{3/2}\left(\sqrt{a^3}W(\phi, T) - 3\sqrt{ak}\right)}{\kappa^2\hbar}\right],$$

$$\psi_6(\phi, T, a) = a^{-\frac{9}{8\hbar}} \exp\left[-\frac{a^{3/2}\left(\sqrt{a^3}W(\phi, T) - 3\sqrt{ak}\right)}{\kappa^2\hbar}\right].$$

6. Conclusions

We have constructed a supersymmetryc Lagrangian for the complete supersymmetric cosmological model of the closed string tachyon in a FRW background, in the 'new'-Θ variables formalism. In order to quantize this theory and extract information of the wave function of the universe, we computed the Hamiltonian of this theory, upon a well suited representation of the Grassmann variables as linear combinations of Dirac matrices. We have found a set of first order partial differential equations for the components of the wave function by means of the supersymmetry algebra. This solution is dependent on the superpotential of the theory, as well as on the scale factor. It would be interesting to study the effects of tachyon in other cosmological models, following the lines of these work.

Acknowledgments

We thank VIEP-BUAP and PIFI-SEP for the support, V. Vázquez-Báez also thanks CONACyT for the studies grant during this work and the organizers of the conference for supporting his attendance.

References

[1] Sen A 2005 *Int. J. Mod. Phys.* A**20** 5513
Sen A 2005 *Phys. Scripta* **T117** 70-75 (*Preprint* hep-th/0312153)
Sen A 2003 *Int. J. Mod. Phys.* A **18**, 4869
Sen A 2002 *Mod. Phys. Lett.* A **17**, 1797 (2002);
Sen A 2002 *J. High Energy Phys.* JHEP**0207**065;
Sen A 2002 *J. High Energy Phys.* JHEP**0204**048
Sen A 2000 *J. High Energy Phys.* JHEP**003**002

[2] Vafa C 2001 Mirror symmetry and closed string tachyon condensation *Preprint* hep-th/0111051

[3] García-Jiménez G, Ramírez C and Vázquez-Báez V 2014 *Phys. Rev.* D **89** 043501

[4] D'Eath P D 1996 *Supersymmetric quantum cosmology* (Cambridge,UK : Cambridge University Press)

[5] Moniz P V 2010 *Quantum Cosmology - The Supersymmetric Perspective* vol 2 Advanced Topis Lect. Notes Phys. 804 (Berlin: Springer)

[6] Obregón O, Rosales J J, and Tkach V I 1996 *Phys. Rev.* D **53** R1750

[7] Tkach V I, Obregón O, Rosales J J 1997 *Class. Quant. Grav.* **14** 339

[8] Wess J and Bagger J 1992 *Supersymmetry and Supergravity* (Princeton, New Jersey: Princeton University Press)

[9] C. Ramírez 1985 *Ann. Phys.* **186** 43

[10] Yang H and Zwiebach B 2005 *J. High Energy Phys.* JHEP**0508**046

[11] Salam A and Strathdee J 1974 *Nucl. Phys.* **B76**

[12] Vázquez-Báez V and Ramírez C 2013 *AIP Conf. Proc.* **1548** 227 (2013) 9th Mexican School on Gravitational Physics: Cosmology for the 21st Century 3-7 December 2012 (Puerto Vallarta: Jalisco México)

AUTHOR INDEX